Media and Water

Media and Water

Communication, Culture and Perception

Joanne Garde-Hansen

BLOOMSBURY ACADEMIC
LONDON • NEW YORK • OXFORD • NEW DELHI • SYDNEY

BLOOMSBURY ACADEMIC
Bloomsbury Publishing Plc
50 Bedford Square, London, WC1B 3DP, UK
1385 Broadway, New York, NY 10018, USA
29 Earlsfort Terrace, Dublin 2, Ireland

BLOOMSBURY, BLOOMSBURY ACADEMIC and the Diana logo
are trademarks of Bloomsbury Publishing Plc

First published in Great Britain 2021
This paperback edition published in 2022

A catalogue record for this book is available from the British Library.

A catalog record for this book is available from the Library of Congress.

ISBN: HB: 978-1-7883-1165-6
         PB: 978-0-7556-3692-1
       ePDF: 978-1-7883-1777-1
      eBook: 978-1-7883-1776-4

Typeset by Integra Software Services Pvt. Ltd.

To find out more about our authors and books visit www.bloomsbury.com
and sign up for our newsletters

I dedicate this book to my mother Joan Margaret Pearson.

Contents

List of illustrations   viii

Acknowledgements   ix

Introduction: Why media and water?   1

Part 1 Communication

1   Media templates for representing water   23

2   Deluge and tempest in the BBC archives   39

3   Socially mediating water for digital hydro-citizenship   57

Part 2 Culture

4   Story-ing water: Liquidity, bubbles, storage   75

5   Remembering and re-mediating women in drought   93

6   Forgetting water: Developing a flood memory app   111

Part 3 Perception

7   The *cultural value* of water and water's impact on *cultural values*   127

8   Riparian media for marginal communities   141

9   Waterproofing media and memory for flood risk   155

Conclusion   171

References   175

Index   201

Illustrations

2.1 (a + b) Photographs: (above) St Peter and St Paul Church, Upton-on-Severn. (below) Boathouse on the River Severn. *Source:* Andrew Holmes   45

2.2 Photograph of the author taking photos of 1947 newspapers with a community group sharing memories of 2007 floods. *Source:* Andrew Holmes   46

2.3 (a+b) Photographs of two sources at Gloucestershire Heritage Hub (above) Folder of 1947 and 2007 print newspaper materials which remains on open display at the hub. (below) Photograph of *The Citizen* newspaper headline from March 14th 1947. *Source:* Andrew Holmes   47

4.1 Bookshop display, Leamington Spa, December 2018   80

4.2 Image of Bubbles of Water Research   85

4.3 Still of home video of River Severn rushing through a resident's home during the 2007 Summer Floods. Image courtesy of the resident who took part on the SFM project   88

6.1 Unofficial flood marks of residents inside and outside their homes and gardens. *Source:* Sustainable Flood Memories team   115

6.2 Screenshot to beta version of the flood app welcome page which uses Open Street Map and Ushahidi platform   117

8.1 The new sign at Ladywell fields drawing attention the regeneration of the river. Photo taken 2016. Courtesy of Andrew Holmes   146

8.2 A Riparian Media and Culture Bubbles Method for civic engagement in river research   154

9.1 The departures entrance of London Heathrow Airport, November 2018. Digital advertisement for Hiscox above the security gate. Photograph by author   160

Acknowledgements

This book is based on a wide range of interdisciplinary research funded by UK, European and Brazilian funding agencies: the Economic and Social Research Council (ESRC) and the Arts and Humanities Research Council (AHRC), as well as the Natural Environment Research Council (NERC), the Federal State of São Paulo (FAPESP) and the Belmont Forum. Some sections of chapter five are based on a blog posting I published entitled 'The Heart and Art of Drought: 10 Things I Have Learned from Brazil' *Drought Risk and You* blog (www.dryproject.co.uk 28 September 2017) and some sections of chapters 2 and 6 are based on the research I undertook for two chapters I co-wrote: one for *Social Memory Technology: Theory, Practice, Action* (2016) by Worcman and Garde-Hansen and the other 'Towards a memo-techno-ecology: mediating memories of extreme flooding in resilient communities' (2016) by Garde-Hansen, McEwen and Jones in Hajek, Lohmeier and Pentzold's *Social Memory in a Mediated World: Remembering in Troubled Times*. The author wishes to thank research participants, research collaborators and colleagues at the Universities of Gloucestershire, Warwick, West of England, in São Paulo and Belo Horizonte for their many research-led discussions. I wish to acknowledge the impact of Lindsey McEwen on my research, she has been a joy to collaborate and write with; and I value the contributions of my colleagues Owain Jones, Andrew Holmes and Franz Krause to the wider *Sustainable Flood Memories* and *Drought Risk and You* projects, both of which inspired deeper research of media archives, media representations and media practices for chapters 2, 5 and 6. The Belmont Forum (ESRC/FAPESP/DFG) *Waterproofing Data* project (led by João Porto de Albuquerque) and the Fapesp/Warwick *Narratives of Water* project with Danilo Rothberg were key to the transnational research with Brazil and Germany and provided the inspiration for chapters 3, 8 and 9. I wish to thank key people for their conversations and collaboration during the preparation of this book: Jacqui Cotton, Nerea Calvillo, Damian Crilly, Carlos Falci, Anna Reading, Gilson Schwartz, Karen Worcman, Mike Wilson, Emma Bee, Peter Kraftl, Carolin Klonner, Nicolas Salazar Sutil, Stefanie Trümper, the BBC archivist Mark Macey, staff at Gloucestershire Archives and the anonymous peer reviewer of the final manuscript.

Introduction: Why media and water?

Water is culture. Water is cultural. Culture is imagined as watery, fluid, droughty, flowing, immersive, trickling and increasingly stormy. It is difficult to separate what is commonly termed 'the natural world' of rivers, seas, lakes and so-called acts of God of flood, drought and extreme wet weather from media representations, social mediation and commercial media production cultures of digitized communication.[1] On the one hand, water affords opportunities for culture to emerge (flood heritage) as water cultures create visions for the future, drawing on water memories and imagined watery (or dry) scenarios. On the other hand, water is conceptualized, represented and socially mediated within contexts of cultural and media production as good or bad seas, watery or dry river basins, beautiful or dangerous coasts, riparian edge-lands, new urban water features and classy town spas. Water is imagined as in abundance or lack, connected to a broader understanding of a watery or dry *sense of place*, with a need to be treated, measured, managed, coerced, respected, modelled and exploited. The latter assumes a representational or socio-scientific framing of water and much has been made of analyses of water that demonstrate water's social power (see Gandy 2014).[2]

However, shifts in geographic disciplines towards non-representational theory – an approach that emphasizes flow, fluidity, mobility, emotion, affect and sensory experiences – have dovetailed with many of the concerns of the arts and humanities (see Abram 1996; Thrift 2000; Cresswell 2012). In particular, the strong emotions and experiences (such as grief, loss, denial, nostalgia, love, refusal, loyalty, anger, violence, joy, awe, tragedy and opportunity) make 'loving the mess' (Kenter et al. 2019) of water stories the starting point of any research project on media and water.

Such blurring of the boundaries between human and non-human, between bodies, languages, texts, experiences and environments has been, in fact, at the heart of media, cultural and communication studies for quite some time. Particularly in British academia wherein media studies draws upon a family tree of French critical

---

[1] In our petro-capitalist conditions the means of mediating and communicating global, national, local and personal water issues through resource-intensive media technologies also contribute to emissions if those media industries pollute, create waste and demand high energy.

[2] The work of Eric Swyngedouw has been instrumental in this regard, in particular his excellent research in *Liquid Power: Contested Hydro-Modernities in Twentieth-Century Spain* (2015: 5) which captures 'stories' of water that 'move beyond H20, the stuff that comes out of taps and irrigation spigots' and focuses on the 'often-invisible actors and agents assembled in and through the flows of water and their benign and erratic behaviour'. For a fictional account of early mid-twentieth-century water issues in Spain read *Drought: A Novel* (2015) by Ronald Fraser.

and cultural theory, the Frankfurt School, the Glasgow Media Group, American mass communication studies, the legacy of Stuart Hall, postmodernism, feminism, postcolonialism, digital humanities, memory studies, and audience and fan studies to name the most salient influences. Yet rarely have water, water issues and water events been addressed in media studies research; perhaps in literature, philosophy and the creative arts of painting, writing and theatre but not in the broadcast, commercial and social media research domains where industry, infrastructure, policy, production and audience are so important. Thus, to return to the representational is to acknowledge that thinking about, measuring, managing, mitigating and adapting to watery issues are as much cultural work as they are political or scientific (involving stories, images, conversation, debate, community, creativity and feelings).

In this book, I cover diverse mediations of water by creative, commercial, publicly funded organizations, broadcasters, businesses and platforms who have been engaged in increasingly connected, connective and affective media performances, practices, texts, images and audience engagements. If the emerging creative industrial revolution of digital and media-tech is also to be a Green Revolution wherein climate issues move from story to action, then media and creative industries need to engage with water as more than corporate social responsibility. Moreover, the messaging of the water industry should see media and culture as more than a storytelling service. My research has been concerned with media and cultural practices that involve reactivating media archives of flood and drought, exploring the creation of new water content for marketing and communications, remembering and forgetting water cultures in heritage management, socially mediating water experiences in everyday life, and engaging communities in water data through visualization, experimentation and sociocultural mapping.

Therefore, this book presupposes a reader interested in water research who knows of the 'cultural turn' in human geography (see Thrift 2000) but perhaps knows less of the various 'turns' in the arts and humanities. Since the 1990s we have seen the significant impact of media, globalization, technology and creative practice on the affective entanglements of humans, non-humans and the natural world within cultural geography. Owain Jones argues in 'After Nature: Entangled Worlds' that we can no longer police the 'arbitrary and inevitably leaky boundaries' between natural and artificial and that the 'nature/culture dualism breaks down discussion into issues of scientific fact and risk, on the one hand, and political desirability or consumer choice on the other' (2009: 295). Yet only recently has media studies (sitting across the arts, humanities, social sciences and technology studies) drawn the 'environment' and the 'natural world' into a deeper questioning of how much media and entertainment are enough; how much energy, resources and labour do media and creative technologies require; and how much media waste is tolerable to consumers who love the arts and culture.

Thus, while the 'cultural turn' has been an ongoing focus, media researchers have been undergoing their own 'turns' towards audiences, policies, technologies, spaces, industries and more recently the Anthropocene, media research has become interested in ecologies and topologies of production and industry. Always critically engaged with practices, places, consumers and experiences, exploring media cultures has meant appreciating media as massive, mundane, mobile, viral, affective as well as defined as flow, channels, peaks, waves, immersion and streaming (all watery atmospheric metaphors). While there has not been much of a 'watery turn' in media studies, concerned as it has been with politics, identity and power, this book contributes to an overall 'greening' (or even 'blue-ing') of media studies more generally that has seen the increased importance of not 'who' or 'when' but 'where' is media and the audience in a changing natural environment?[3] It is the intimate entanglement of media (past and present) with our watery environments that this book seeks to explore so that policymakers will increasingly expect to have arts and humanities researchers involved in discussions about managing water, making room for rivers, tolerating drought, adapting to flood, dealing with pollution and engaging the public in water issues.

Consequently, in media, communication and cultural research, a natural resource such as water becomes as much a question of managing stories, media, narratives, mythologies, representations and social experiences of audiences and storytellers as it does about managing water in a changing climate by public and private enterprises. While human geographers hopefully find the depth and breadth of media research in this book illuminating, I am really speaking to the arts and humanities research community. Not only the 'creative arts' (performance, theatre, writing) through which art, craft and creativity have been embraced by a wide range of non-representational theorists in geography (see Boyd and Edwards 2019) but mainly those in media studies wherein art, culture, creativity are not easily separated from marketing, commerce, corporations, globalization, technology and business interests.

Researchers of media production, distribution and technology, media texts, cultures and audiences as well as digital media, social media and participatory media are entangled with water without really knowing it. From the impact of popular television productions that increase tourism to water-stretched regions (as in the BBC series *Poldark* and its effect on Cornwall, UK) to climate change protest on plastic pollution (the impact of *Blue Planet*, for example) to the growth of creative work on flooding (see Bryan and Mary Talbot's graphic novel *Rain*, 2019, set in a flooded North of England). From ignoring media histories of representations of flood, drought and extreme weather to leaving the communication of water scarcity to the science community and privatized water companies. From not questioning in enough depth the policy narratives that mythologize water within discourses of hope and optimism, fertility and vigour expressed in terms of connection, newness, health, flow, strength, social cleansing and well-being,[4] media and cultural researchers have missed altogether the increasing dryness of all our environments. If we have bothered to pay attention to water, then only 'biblical' flooding provides the memorable media templates worth

---

[3] Western media scholars cannot ignore environmental change and particularly flooding, extreme weather and drought considering the global and social mediatization of events such as Hurricane Katrina (2005) and Hurricane Harvey (2017) in the United States, Storm Chiara and Storm Dennis in the UK (2020) and the increase in heatwaves across Europe. The implication of media production processes in anthropogenic climate change is being addressed by the AHRC funded *Global Green Media Production Network* established at the University of Warwick (2019–20).

[4] 'We believe a deeper connection to local water bodies can bring a new cycle of community hope and energy that will lead to healthier urban waters, improved public health, strengthened local businesses, and new jobs, as well as expanded educational, recreational, housing, and social opportunities' (Urban Waters Federal Partnership 2011, USA: online). Or see Myers et al. (2012) 'A public health frame arouses hopeful emotions about climate change.'

analysing. Orlove and Caton (2010: 404) have written that 'value, equity, governance, politics, and knowledge' are key to understanding 'waterworlds', but in this book I make the case that so too are culture, media and communication.[5] Research from my own field contributes to understanding the construction and experience of many diverse and hidden waterworlds that shift in the light of broader sociopolitical issues, such as in the case for the Mediterranean Sea, reframed in recent years from idyllic holiday destination to human migrant disaster zone.

It is possible, then, to see in water cultures 'manifestations of optimism' as 'a common *mode* of viewing the future [even if that is a memorial message to that future about taking more responsibility for water], whilst at the same time acknowledging the conflicting variety of its expressions' to rework Bennett's *Cultures of Optimism* (2014: 168). For example, why would a privatized water company refrain from socially mediating messages of watersaving during a prolonged period of dry weather and keep its 'drought plan' in the backroom of its website? Why does a community remove flood marks from its neighbourhood of newly built homes on a flood plain and lament iconic media images of a flooded historic building as fixing that building under water? How and in what ways do we remember great droughts (if at all in the context of spectacular heatwaves and fire weather) and what do we choose to forget of water scarcity? Close attention to mediated domains of water is so important: education, the family home, the water industry, the environmental agencies, local and national governments, NGOs, religions and social movements, and as water experts become media producers, the science domain itself emerges as offering another mode of mediating the past, present and future of water. All of these create multiple *cultures of water* (potentially translocally connected by media cultures) in which institutions, organizations, individuals, collectives, communities, professionals, stakeholders, users, audiences, consumers, taxpayers, believers play out their roles and seek to shape (or are shaped by) an overarching *culture of water* in which water remains the start and end point for discussions about human life, rights and happiness.[6] Such an opening to a book on media and water provides the basis for exploring a *cultural policy of water*[7] or, more interestingly, the possibility of *water as a form of cultural policy making*,[8] wherein 'cultural policy can be understood in its broadest sense as encompassing any course of action that is deliberately designed to shape attitudes, values and behaviours' (Bennett 2014: 172).[9]

---

[5]   The arguments concerning how we measure cultural value share criticisms with how nature is valued and the evidence of value upon which 'waterworlds' are judged. For more on nature, culture, value and ecosystems, see Fish et al. (2016) in 'Conceptualising cultural ecosystem services: A novel framework for research and critical engagement.'

[6]   Scan the 50+ items on the European Environment Agency's webpage from 2001 to 2018 and we can see engagement with communities, water and video photo competitions, and an ecosystems approach. Available at https://www.eea.europa.eu/themes/water/highlights. Accessed 1 December 2018.

[7]   Using arts, media and culture to shape attitudes to and expressions of water cultures.

[8]   Acknowledging water as a non-human actor in the shaping of human cultures.

[9]   Bennett states cultural policy 'is often represented as a relatively minor domain of government, associated mainly with the support or regulation of the arts and media' while 'in the broader sense of shaping attitudes, values and behaviours, then it not only takes on a much greater significance but it can also be said to have existed for centuries, even though it may never have been described as "cultural policy"' (Bennett 2014: 173).

'Water' is, as a concept and topic, very much embedded in the science domain – historically produced through the hegemony of chemistry according to Hamlin (1990). It is also a resource for 'citizenship' and for policymakers as in the European Union Water Framework Directive (2013). Water is also continuously reimagined by artists and creatives who draw on a long and complex history of the representation and use of water in art, such as in the mock print advertising by Onesto (2015) during a Californian drought.[10] Cultural policy has had an impact on water's place within human environments. It is a product of cultural production in leisure, tourism and heritage industries; there are museums of water, steam, seafaring and seascapes; memorials to floods and storms, tempests and hurricanes; areas of outstanding natural beauty portray and market their water parks and wetland reserves; and everyday water experiences in swimming pools, thermal spas and lidos alongside the consumption of successfully marketed global drinks all commodify and contain water, packaging it within concrete, tiles, glass and plastic.

Yet the study of 'water' as a core disciplinary area has largely remained in the sciences and social sciences: chemistry, hydrology, geography, engineering, biology, economics, computational modelling within ecology, alongside sustainability and risk research in climate adaptation and resilience. We do have academic histories of water, such as in *A History of Water in Modern England and Wales* (1998) by John Hassan or fresh perspectives on water in the context of power and politics, such as in David Lewis Feldman's *Water* (2012) or Jamie Linton's (2010) *What Is Water? The History of a Modern Abstraction*, but these tend towards economic and social histories or offer environmental policy perspectives. While most publications are increasingly concerned with water conflicts, risks, controls, regulations and management, there is recent research that considers more deeply water and its meanings, symbolism, spirituality and its relationship to humankind, memory, representation, as well as its connection to cultural and creative practices.[11] Yet much of this literature has not drawn upon scholarship from media, culture and communication studies, which is an area of research that has really only touched on water (such as extreme weather,[12] flooding and drought) through analysing newsworthy mediated events in the context of climate-related news values. For example, the Tsunami of 2004 (see Hastrup 2008), Hurricane Katrina in 2005 (see Littlefield and Quennette 2007; Robinson 2009), the Pakistan Floods of 2010 (see Murthy and Longwell 2013), the Brisbane Floods of 2011 (see Bohensky and Leitch 2013) and the representation floods from the 1950s to 2000s

---

[10] Chris Onesto (2015) 'California Drought' https://www.kcet.org/shows/artbound/art-water.

[11] Connectivity is at the heart of *The Power and the Water* project 2013–16: Arts and Humanities Research Council (AHRC) funded (led by Georgina Endfield, Peter Coates and Paul Warde). Available at http://powerwaterproject.net/. Many of the publications and papers from the project touch on cultural heritage but none on media and communications.

[12] Extreme weather is not the focus of this book but has been addressed by the 2013–17 AHRC project *Spaces of Experience and Horizons of Expectation: The Implications of Extreme Weather Events, Past, Present and Future.* The production of the TEMPEST database states it 'will be a freely accessible and user friendly resource on the UK's climate history. It will include records of all types of "extreme" weather and will be searchable by weather type and location, as well as by keyword. Some records will also include digitized images of the original sources and links to supporting audio or video files'. Available at https://www.nottingham.ac.uk/research/groups/weather-extremes/research/tempest-database.aspx. Accessed 1 December 2018.

in the UK (see Furedi 2007; Escobar and Demeritt 2014). These may offer very different perspectives on water as a result of the climatic and geographic distinctions between regions and weather systems, but they are also different socially and culturally due to the 'dynamic inheritance of social attitudes towards water embedded in place and context while travelling across settings' (Garde-Hansen et al. 2017: 385).[13]

Thus, in the last decade or so, academic studies from arts and humanities perspectives have begun to explore the cultural histories and representations of water, the cultural production of water and the impact of sociocultural practices upon water, rivers, seas and climate. For example, within heritage studies we find analyses of museums and the role of water in the slave trade or shipwreck commemoration (see Rice 2012; Pearce 2018); the critique of coastal heritage-making in Jones and Selwood (2012) and Setten (2012); or an analysis of 'dockside kitsch' landscapes in Atkinson (2008). Within literary studies, an analysis of medieval poetry and flooding in Griffiths and Salisbury (2013) drew together rivers with grief, which connects more recently with accounts of flooding in commemorative culture (see Hall 2018). Water is often analysed for its symbolic meanings rather than as a cultural agent or as a non-human actor. Eco-critical readings of the flooded Fenlands in Graham Swift's 1983 novel *Waterland*, as in Head's analysis for *The Green Studies Reader* (2004), lay some of the foundations for a recent transition from 'greening' the arts and humanities to a 'Blue Cultural Studies' as suggested by Mentz (2009).[14] The latter's reading of the sea in Early Modern English Literature inspires a blue cultural studies approach to be taken further, perhaps by Picken and Ferguson on the promise of a 'blue planet' (2014) or in Elspeth Probyn's (2016) *Eating the Ocean*.[15]

Thus, more than simply a backdrop, the role of water in literary and cultural studies, such as in Cohen's *The Novel and the Sea* (2010) or in cultural texts explored in Charlotte Mathieson's edited collection *Sea Narratives* (2016), is indicative of what Pearce defines as a '"maritime/coastal turn" in the social sciences and humanities' (2018: 55). This can be found in UK research projects funded by the AHRC such as Holdsworth and Penny's amateur dramatics in the British Royal Navy (AHRC 2017) or in the University of Nottingham's *Rising from the Depths: Utilising Marine Cultural Heritage in East Africa* (RCUK 2018). The emergence of what we might call a pluvial-fluvial-riverine turn found in McEwen's *Living Flood Histories Network* (AHRC 2012) or Jones's *Hydro-citizenship Project* (AHRC 2014–2017)[16] has informed the research

---

[13] Despite their differences one might place these in the context of a globalized mediatization of extreme weather events with a narrative arc – climate uncertainty and change – that is contributing to a global watery imaginary and a re-examination of economic, political and social histories in the light of histories of water and human habitation.

[14] During the writing of this book a *Blue Humanities Network* was established at the University of Warwick, UK, in July 2019. On 29 October 2018 Prof Susannah Radstone gave her Memory Lecture to the Warwick Memory Group entitled 'Hard Landings: Memory, Place and Migration' focusing on St Kilda, New Zealand, in the context of watery migrations. The growth in water/memory research is evidenced by the 2017 conference *Oceanic Memory: Islands, Ecologies, Peoples* in Christchurch, Aotearoa/New Zealand, which covered literature, film, performance, television, creative writing and community arts.

[15] See *Symploke* journal volume 26 on 'Oceania in Theory' and the 2019 Call for Papers for volume 27 'Blue Humanities'.

[16] See the *Hydrocitizenship* project (2014–17) https://www.hydrocitizenship.com/ and also the more sprawling online community http://www.hydrocitizens.com/home.

that underpins this book and we can see water moving within new disciplines and across disciplines. Pearce, citing Lambert et al.'s (2006: 479) positioning of 'seas and oceans at the centre of [academic] concerns', suggests we should notice in the arts and humanities a shift away from the terrestrial (2018: 55), and this is signalling more than a recognition of a mostly blue planet.[17] It suggests, I will argue, in the context of this book the emergence of some new political and cultural economy frameworks for analysing the transcultural, trans-medial and transdisciplinary values of water; a decolonizing of the academic curriculum; and, a greening of media practices, cultural policy and heritage management (in the light of climate crisis) so as to (re)value water histories and futures.[18]

Therefore, for these new movements of water research to be more fully understood in the context of a history of developing mass media, global communication networks, and increased and democratized access to creative and media technologies, we are required to rethink the epistemological frameworks upon which our knowledge of water has been constructed. Once we circulate water research through international partnerships, researcher exchanges and interdisciplinary knowledge construction, we find that water research will have little impact if it is not deeply connected to organizations, the private sector, communities and stakeholders that deal with and use media every day. This requires a fluid and connective way of working so that media, communication and digital cultures become infrastructural requirements for the connectivities and circulations of water research and for new appreciations of water cultures to move more freely around the world.

These new knowledge frameworks may also be more 'wet-and-dry' as in Krause's (2017) 'amphibious anthropology' that considers 'hydrosocial' relations of volatility, creativity and rhythm.[19] They may be epistemologically in 'flux' as in Chen, MacLeod and Neimanis's collection *Thinking with Water* (2013), in which the preposition 'with' indicates the mutually constitutive approach of the many chapters that consider how to think 'with' water rather than about, on, upon or against. Citing Bachelard's (1983 [1942]) *Water and Dreams: An Essay on the Imagination of Matter*, they introduce the 'material metaphors' of flux,[20] flow[21] and circulation in how we theorize, as a code for thought, emotion and memory:

---

[17] Future study needs to explore how 'race' is expressed in this new research area, and it is worth cross-referencing the recent expression of a 'blue planet' with issues of race and migration and racist fears of a 'black planet'; as well as, the framing of channels and seas as migrant routes wherein these waterways are recoded as 'bad'. This would be quite a different framing from Paul Gilroy's 'intercultural positionality' of the 'ocean' in his important work *The Black Atlantic: Modernity and Double Consciousness* (1993: 6).

[18] This must be balanced with the impact of media on environments (see Cubitt 2016) albeit most research in this area is focused on terrestrial impacts (e.g. mining for minerals).

[19] See the seminar 'A Dying Lake, a Living Sea, and Other Bodies of Water' by Jennifer Lee Johnson, Assistant Professor, Purdue University, Department of Anthropology at Aarhus University, 4 April 2018.

[20] John Urry argues that: 'Flux involves tension, struggle and conflict, a dialectic of technology and social life [...] the complex intersections of immobilities and mobilities' (2007: 25) and Thussu (2007) focused on media flow and contra-flow.

[21] See Raymond Williams https://en.wikipedia.org/wiki/Flow_(television) and O'Sullivan (1998) 'Nostalgia, Revelation and Intimacy: Tendencies in the Flow of Modern Popular Television'.

In everyday speech, emotions 'flood', 'bubble up', and 'surge'; a 'dry' text is one that lacks feeling and passion. We 'freeze up' with stage fright, join or diverge from 'mainstream' populations. Money 'circulates'; commodities 'flood' the market. The past is a 'depth' and time 'evaporates'. Neither is the realm of theory immune to inspiration from the liquid world: aqueous dynamics of 'flux' and 'flow' characterize qualities of indeterminacy and continuous change within many contemporary epistemologies, whole feminist concepts of 'leakiness' and 'seepage' have been mobilized to identify crucial porosities in bodies and theories alike. [...] Just as water animated our bodies and economies, so it also permeates the ways we think.

(Chen et al: 10)

Thus, thinking hydro-symbolically means addressing issues such as class, race, gender and social exclusion and the impact these have on water and vice versa. While this book, *Media and Water*, will make some inroads into these important issues, it remains for future researchers to deepen the intersections within specific media forms such as cinema or cultural forms such as museums, or with regards to specific intersectional experiences. Initially, there is a job of work to be done on positioning media studies of water as central not only to the discipline of media, communication and cultural policy research but also to science, science communication and water policymaking. For water, water issues and watery events are no longer hidden, peripheral or one-off; they are consistently, pervasively present in contemporary media culture.[22]

There are many studies in the social sciences and business that have focused on water's use values in an instrumentalist and rationalist mode alongside and integrated with social and cultural factors (see Swyngedouw 2009). If water is not managed well then water, like the culture itself, will be unevenly distributed for if 'particular trajectories of socio-environmental change undermine the stability or coherence of some social groups or environments, while the sustainability of others elsewhere might be enhanced', then conflict occurs. 'Consider', says Swyngedouw (2009: 57) by way of example

how the provision of water to large cities often implies carrying water over long distances from other places or regions. The mobilization of water for different uses in different places is a conflict-ridden process and each techno-social system for organizing the flow and transformation of water (through dams, canals, pipes, and the like) shows how social power is distributed in a given society.

This has produced a normalization of uneven geographic experiences of water and, in particular, flooding which in some contexts disappears from view because it is happening so frequently.[23]

---

[22] A recent example in the British press would be UK flood and storm coverage of February 2020, in which repeated floods in the North of England are represented as the new (temporal) normal. A poignant statement from a 64-year-old householder bears this out next to her photograph: 'Barbara Campion outside her home. She and her neighbours gave up on carpets several floods ago' (Pidd, *The Guardian*, 22 February 2020).

[23] A sign of the disruptive flood event becoming normalized is the reframing of flood victims as a becoming part of a flood community such as in the emergence of Facebook groups of 'floodies'.

Thus, if water can become emptied of its contradictory symbolic and spiritual meanings within highly developed societies and lose its capacity to remind humans of its longer cultural and social storytelling, then it is simply a commodity to be circulated in technological infrastructures (much like television texts in an online platform ecosystem). This dichotomization of our knowledge base of water between the arts (such as ethics, philosophy, literature and religious studies) and the 'harder' sciences (of engineering, modelling, management and data analytics) has not been particularly helpful in bringing researchers together to address deep and long-standing issues of scarcity, flood and storm, pollution and drought. One of the general propositions of this book is that the arts and humanities (i.e. artists and creatives) bring skillsets to activate people (in fast ways), and the arts, in particular, have the capacity to break down hierarchies and address those 'under-served audiences' through both proximate and distant communication techniques. These researchers (when brought together) also have a deep knowledge of disconnected archives, memories and heritage (monuments, texts, images, artworks, media, performances, local and national records) that can be connected with anthropological, geographic and geological records, and archives. Such archival connectivity could save significant time in developing responses to climate uncertainty.

In a changeable climate, issues of water really do need a collaborative and strongly interdisciplinary response if the extremes of human experience and adaptation, and the challenges around stressed natural resources, are to be brought together into a constructive dialogue.[24] This book argues that the mediation of water by creative, media, cultural and communication industries, practices and citizens plays a strong and important role in not only thinking with and without water, but in feeling, sensing, remembering and imagining water as a non-human actor in a highly mediated environment. Science communication, government policymaking, mainstream media, local storytelling, everyday experiences, personal attitudes, social movements and family values all inform pro-environmental becoming ('green' and 'blue' planet attitudes), with individuals and their behaviours rubbing up against and within structures and organizations (where groupthink can be just a prevalent as independent voices). These are mediated by technologies of culture, creativity, communication, broadcasting and networking, and these technologies are unevenly distributed and controlled. This does not mean this book is not critical of media representations, cultural production or communication networks, but rather it considers the affordances of these for circulating water stories.

While this book is not focused squarely on science communication and environmental policy communication from government and media organizations (other texts have covered this far more thoroughly, such as Sheppard 2005; Lowe et al. 2006; O'Neill and Nicholson-Cole 2009; O'Neill 2013), it does recognize that these are not going to be effective unless eco-consciousness is not already well understood through reference to inherited, diverse, disaffected and resistant

---

[24] See the work of Sam Illingworth; https://www.samillingworth.com/academia explores how academics are working against their own expertise and addressing underserved audiences through creativity.

cultural values of water. Public engagement is far more than nudging behaviour and corralling consent; it is storytelling, memory, values and identity recognition, all of which take time, deep listening and respect but can speed up adaptation if taken seriously. Audiences need to be recognized as users, readers, viewers, creatives and media literate producers already primed to use narrative-building and storytelling techniques through a wide range of media and cultural activities. The basis of this proposition is built upon the idea that water users are entangled with multiple scales of water stories from the macro-climate-related narratives to the everyday uses of the hosepipe. Audience members may shed tears over a YouTube video of plastic straws stuck in the nostril of a sea turtle and may never purchase a plastic straw again, but they might also continue to wash their cars in hot weather regardless of the messaging from their local water company.

Therefore, the Sciences and the Humanities need to recognize our own entanglement with one another, as Karen Barard has argued in her work on 'new materialism':

> There was the notion that what is needed is a synthesis; a synthesis or a joining of the Humanities and the Sciences as if they were always already separate rather than always already entangled. So that there would be Science with matters of fact, and nature, and so on, on one side, and Humanities, meaning, values, and culture, on the other, and somehow that there would be a joining of the two. So, we talked about the ways in which there are entanglements that already exist between the Humanities and the Sciences; they have not grown up separately from one another.
> (cited in Dolphijn and van der Tuin 2013: 66)

Recognizing and daylighting the existing and long-standing cultural narratives and experiences of a watery sense of place that are entangled with engineering, forecasting, scientific histories and futures need to be undertaken, administrated and shared in ways that are respectful to and of the communities served by and who serve the rivers, lakes and coasts while facing changeable climates. Thus, I will argue throughout this book, offering case studies that explore media histories, presents and futures, that we need more studies of the cultural, social and digital mediations of water, such as Hall's (2018) analysis of local and regional newspapers in exploring the 1953 East Coast Flood aftermath or Morgan's (2018) use of newspapers to reinterpret the cultural memory of the 1914 Australian drought. For these reasons, *Media and Water* is an intervention in this direction, and it starts with water as culture and water as cultural because of a limiting assumption that media are there to simply represent, send and share messages about water, and not much else.

Past waters and the cancellation of water's future

Whether the future emplotted is (post-)apocalyptic and characterized by socio-economic and ecological collapse and species extinction, or one of resilience, adaptability, and sustainability, or somewhere in between, such fictions stage cultural memories of the Anthropocene and so an aetiology of the conditions that are imagined in the future but which are unfolding in the present of this literature's production and consumption.

(Craps et al. 2018: 501)

Craps et al. (2018) address the fluidity of memories for making futures possible in a time of finitude and risk. Memories of water are mediated through increasingly globally connected people with media devices that have the capability to remember or forget the water in and under the cities, streets and land where they live. Media make visible water inequalities even in times of a shared global disaster such as the Covid19 pandemic of 2020 (e.g. flooded communities unable to self-isolate in safe homes in the UK or African-American women and children with water cut off due to non-payment in Detroit, United States, where a basic right to water is absent).[25] Media (like water in rivers, seas and oceans) circulate globally, crossing national boundaries, in spite of efforts to channel, manage and control. Yet water inequalities, water crises and water politics are addressed within national media and public spheres as if water and the meanings it generates must be read separately against the social, cultural, religious, historical, political and economic schema of each nation state. What mechanisms can be produced and what is the role of media for a transnational/trans-setting sharing of stories of water?

My interdisciplinary research has explored the nature of sustainable flood memory, drought narratives and stories/memories of water (i.e. in urban river regeneration contexts) by incorporating a deeper understanding of the relationship between the personal, social and hydrological: that is, in terms of a 'watery sense of place', folk memory of flooding, water heritage, water scarcity stories and river nurture-neglect narratives. The research gathers evidence for how the personal-social-hydrological has been and is materialized in communities, landscapes, in media and in cultural memory (from local archives to water festivals to family anecdotes). It proposes theoretical, methodological and applied ways for thinking through how this media-memory-knowledge can inform strategies for social learning that are vital components of flood-risk planning, water scarcity governance and river regeneration policies and activities. The aim has always been to connect cultures of water use, water history and water well-being/values so as to increase community resilience through deeper analysis of media and culture. More recently, I have been collaborating with researchers in São Paulo and Belo Horizonte, Brazil, because comparing, contrasting and connecting the soft communication and cultural aspects of water governance may identify 'gaps' and afford the use of media, creativity and culture research to explore those gaps further.

The intellectual contribution has been to connect cultures of practice (media/memory and flood/drought/river management) and address these with communities and stakeholders through an understanding of the remembering and forgetting practices of those communities. The impact has been driven by a need to enable more connected media and memory work across scales so as to develop better

---

[25] It's worth noting that Covid19 was described in watery metaphors in 2020 and the writer Damian Barr challenged these on Twitter, 'We are not all in the same boat. We are all in the same storm' and therefore in many different types of boat, some without oars. See Peggy Norman, 23 April 2020, 'What Comes after the Coronavirus Storm?' *The Wall Street Journal.*

understandings of how communities can support their own adaptive capacities in a changing climate, drawing on the past for the future. Follow-on research findings on trialling digital storytelling and 'narratives of drought' have extended media, memory and water scarcity research further into collaboration with significant UK stakeholders such as the National Farmers Union, the Canal and River Trust, the Environment Agency and representatives of the business community.

Yet disciplinary silos continue to exist, and the main challenge in my research field has been to continue to operate with a sense of 'strong' and 'genuine' interdisciplinary work. For example, in my collaboration with the UK's Centre for Ecology and Hydrology (CEH) I have been struck by the 'in the field' methodologies used to understand the impact of factors on the nurture or neglect of selected parts of the natural environment. This is achieved through 'mesocosm research', wherein a part of the 'real environment' is sectioned off and compared to the lab results. In media, memory and cultural studies we could respond to the challenge of interdisciplinarity (connecting cultures of research) by genuinely and strongly sharing methodologies. 'Cultural mesocosm' research is one area to explore as an approach to connecting up the personal-social-hydrological. Likewise, 'telling stories', 'sharing memories' and constructing 'narratives of water' can be another way of 'storying the science'. As water becomes more scarce we are going to need to (a) connect with cultures in the world where such scarcity is undeniable and remembered (and connect across social, religious, cultural and ethnic boundaries); (b) find new ways of connecting the different sectors that are all invested in water (from gardeners to the drinks industry, from golf courses to water companies, from riparian communities to farmers); and (c) consider how we might implement an ongoing cultural policy of water that promotes policy transfer.

Towards a cultural policy approach to water

In the final chapter of Bell and Oakley's *Cultural Policy* (2015) they signal a need for change within studies of cultural policy and a need for cultural policy itself as practised by governments, NGOs, institutions and corporations to move in new directions such that culture is not seen only as the context in which policy acts, but that culture acts on policy and that resources previously understood as only having economic value are born out of their cultural values and creative potential. It is timely, for this book on *Media and Water*, that they should argue:

> Rather than seeing culture as a resource to be used economically, as the creative economy of cultural industries traditions generally do, the argument would be to see 'economic' resources from *water* to housing to green spaces in cultural terms, to help understand what they mean to people and hence how they can be valued in terms other that the economic – or through a radical rewriting of the definition of the economic.

(Bell and Oakley 2015: 157–8, my emphasis)

This is a contested issue for indigenous communities, in particular, or those who build their cultural identities upon articulations of water's intangible heritage or for underserved communities in fear of urban water gentrification or who believe that concrete is the symbol of progress. In Craft's (2017) case study in the edited collection by Thorpe et al. (2017), *Methodological Challenges in Nature-Culture and Environmental History Research*, she addresses the 'water law' of the Anishinaabe First Nation community of Canada and asks the key question concerning the 'murky' and contested claims over water: 'Could [our water law] shift priorities from ownership and the economic value of water to acknowledgement and protection of the spirit of water itself' (Craft 2017: 106)[26] and how might such an approach offer hope to working-class communities of any ethnicity whose local water is polluted?

Such exploratory questions may showcase cultural knowledges at odds with Westernized and technocratic solutions, as Hossain and Marinova (2012) found among the rural people of multicultural Bangladesh, wherein spirituality among the largely uneducated existence of those committed to the Baul tradition demonstrated a well-established blueprint for water management in contested spaces. They argued:

> Irreverence or ignorance of water related spirituality by modern societies is the fundamental reason for scarcity, pollution, over-extraction, mal-utilisation and aggressive politics of water. Values-driven water management is emphasised as the sustainability breakthrough and an essential requirement for proper development.
>
> (Hossain and Marinova 2012: np)

Thus, could such a 'blueprint for sustainable water management' follow the same advice that Hossain and Marinova offer, regardless of local conditions? For such a blueprint needs to be multiscale, multimodal and operate across discourses, forms and practices be they media, education, songs, folk stories, local knowledges, memories, anecdotes, proverbs, poems, performances and much more that falls within the realm of the arts and humanities. Such connectivity across domains also opens up the possibility of new articulations of and for water in European contexts where lay knowledges of water have been neglected and may resurface as new kinds of ecological awareness and where the 'migrant' trope can be extended widely to many cultural identities.[27]

What then are the disciplinary assumptions, the inclusive *exploratory questions* and the necessary *scoping topics* that researchers and non-academic stakeholders need to address if they are to develop a cultural policy approach to water and its management?

---

[26] For more on decolonizing methods, see Joan Cruikshank's (2005), *Do Glaciers Listen? Local Knowledge Colonial Encounters and Social Imagination.*

[27] See, for example, Owain Jones and Katherine Jones (2017), 'On Narrative, Affect and Threatened Ecologies of Tidal Landscape'.

Exploratory questions	Scoping topics
• What are the old and new water knowledges? • How are various actors in various networks integrating these knowledges and through what kinds of established and new media infrastructures?	• How knowledges are built, organized and shared by different social and cultural groups. • The impact of democratization, media, digital and participatory culture. • Include 'non-knowledge', affect, memory, sensation, feeling, the spiritual and emotion.
• What are the settings, opportunities, challenges to engagement in cultural values of water? • How ephemeral are the media, cultural and social networks around water issues and is their learning transient and in danger of being lost?	• Consider the horizontal axis of community and social events, community media and arts, serious games, social movements and ecological solidarity organizations, all of which can be connected and organized across local, national and global scales. • Consider the challenges and opportunities of digital devices, cloud storage, new interfaces and apps, new infrastructures for connecting domains.
• What are the opportunities and challenges of different social and media platforms during water events and in their aftermath for remembering, knowledge gathering, decision-making and action?	• Consider the uses of YouTube, Weibo, Facebook, WhatsApp, WeChat, Flickr, Twitter, Instagram and LinkedIn as ways of connecting across sectors, scales, domains and national borders. • Consider the persistence versus the transience of different media.
• What are the impacts of new and changing media practices on social and cultural learnings about water issues? • How are social media platforms and social networks being used by different interest groups?	• Consider participatory knowledge, co-production, citizen curators and science communicators, edu-gamification, early-warning apps and creative arts approaches to water communication. • Consider developing new integrated digital media infrastructures that draw together civil society, NGOs, businesses, environmental regulators, government, local communities.

I suggest some starting points above which underpin the research this book brings together on mediating water management, flood risk, urban river regeneration and drought memories, filtering these issues through media, communication and cultural studies in order to both widen and deepen the conversation around water in a changing climate.

There is a need for academic research from media, communication and cultural studies to fully enter the water research domain drawing on an integrated science-narrative approach that acknowledges that 'water' (its availability, abundance, scarcity, quality and management) is increasingly mediated by professional communicators and mediatized by creative and sociotechnical devices, apparatus and infrastructures. None of this communication is neutral. Water is in 'media' – represented by television, film, radio and print – and water's actions on human habitats (flooding, drought, extreme weather) produce stories, memories and traumas, offline and online, public and personal. Water produces a sense of place, identity and belonging particularly in countries surrounded by water or with an increasingly dry sense of place.

Furthermore, water's representation in media[28] (from 1947 to 2017[29]) can be researched in ways that reveal critical insights into genre (how we have told and tell the story of water, water as a character, water stakeholders as characters), gender (who tells and gets to tell those stories, roles for women, recuperation of hidden stories of 'women's watery work'), race (how water is represented and channelled in ways that benefit one community and disbenefit another through structural inequalities and groupthink) and generation (why and how one generation tells those stories to another and to what effect, the emergence of a climate change narrative, the role of scientific discourse and the place of folk/local knowledge). Put simply, how was and is water 'story-ed' by media and through media, with what kinds of intention, motivation and story-like properties that produce water stories as a kind of cultural habitat? These issues will be addressed through a combination of literature review and case studies within three sections: communication, culture and perception.

Why mediate water? Why now?

On the one hand, there is much to 'inform' the general reader about water. Owain Jones in his Arts and Humanities Research Council – funded 'Hydrocitizenship project' (2014–17) – calls it a 'water zeitgeist in art, literature and poetry' with over 100 books on the project's wiki: 'Why?' because of 'a reaching out to our immediate ecologies' for 'water is a distinctive force of nature – culturally, politically, ecologically vibrant' (Jones 2018 (online)).[30] The website GoodReads lists over 1,500 popular culture and popular science books concerning 'water'. Most of those in the Top 20 articulate disaster, struggle, disempowerment and fear.[31] There are many fiction and non-fiction books similarly titled: *Water Wars: Coming Conflicts in the Middle East* (1993) by Bulloch and Darwish, *London's Water Wars* (2000) by Graham-Leigh, *Water Wars: Privatization, Pollution and Profit* (2016) by Shiva, *Water Wars: Drought,*

---

[28] One has to place limits around 'media' in terms of the space available in this book. There are many forms and practices of media that this book does not cover in detail, cinema, popular music, animation or computer games, for example. They may be touched upon where relevant but broadly, the book covers broadcast, print media, social media and participatory media cultures as under the category of 'media, communication and culture' research.

[29] One often has to place arbitrary dates around research parameters. I have chosen the seventy-year time-span because it broadly covers the broadcast to post-broadcast era in fully developed media infrastructures such as the UK. The year 1947 also marked a significant Winter flood in the UK, which was an early example of a 'media event' at least in terms of the press and radio. Thus, mediated storms, floods, droughts could become the measure of time in this book.

[30] Jones points out the following books as belonging to this 'water zeitgeist': *Rain: A Natural and Cultural History* (Barnett 2015); *Downstream: A History and Celebration of Swimming the River Thames* (Davies 2015); *A Recipe for Water* (Clarke 2009); *The Water Book* (Jha 2015); *Caught by the River: A Collection of Words on Water* (Barrett and Turner 2009); *The Fish Ladder: A Journey Upstream* (Norbury 2015); *The Last Drop: The Politics of Water* (Gonzales and Yanes 2015); *Waterlog: A Swimmer's Journey through Britain* (Deakin 1999); *The Fabric of Space: Water, Modernity and the Urban Imagination* (Gandy 2014); and *The Sea Inside* (Hoare 2013).

[31] For a more academic perspective on natural disasters see Oliver-Smith (2002) 'Theorizing disasters: Nature, Power and Culture' and G. J. Schenk (2007) 'Historical Disaster Research. State of Research, Concepts, Methods and Case Studies'.

Flood, Folly and the Politics of Thirst (2003) by Ward, *The Water Wars* (2011) by Stracher, *Water Wars: Fight to the Last Drop* (2017) by Whitehead and *The Great Lakes Water Wars* (2018) by Annin. These are connected as a range of books to frame water in terms of conflict, scarcity, drought, rights, power, pollution, privatization and mismanagement. What Howe and Boyer (2020) have termed 'hydrological globalization' as water is redistributed across the planet. They demonstrate that for the 'general reader' this is an area where much fear can be generated: the one resource we cannot live without for more than a few days is presented as always *in jeopardy*, and this becomes a key cultural resource for drama.

There is a need, in the context of all this contemporary jeopardy, for an exploration of the histories of mediations of water that offer alternative visions and invite the reader to critically reflect upon the current remediation of those histories through social and more participatory media across globally networked platforms. When there is crossover between the arts and sciences on water issues, this occurs largely in journal articles or chapters focused on a single water issue. For example, as in 'Public Health in the UK Media: Cognitive Discourse Analysis and Its Application to a Drinking Water Emergency' by Knapton and Rundblad in *Contemporary Critical Discourse Studies* edited by Hart and Cap (2014). More broadly the 'sociality' of water is addressed in terms of climate change and ecology, such as in 'Social Limitation of Sustainable Water Consumption' by Glasauer in *Urban Ecology* edited by Breuste et al. (1998). Neither of these really gets to the heart of a media, cultural and communication approach to water that would draw attention to the discourses, forms, practice and technologies that assemble to communicate water in a digital age.

Importantly, the business and NGO activity of communicating water are well covered in the edited collection *Water Communication: Analysis of Strategies and Campaigns from the Water Sector* edited by Hervé-Bazin (2014). While this collection takes a broad approach in that it is less concerned with one single geographic context, it does acknowledge that water circulates around the world, water is inside people as well as places, and it should be a 'right' for all in terms of addressing many different mass communication and marketing strategies. While water is a universal and global issue, and media have become globally connected phenomena, there remains a need for deeper analysis of the place, culture and communication specifics of mediating water across connected experiences of flood, drought and water management. Tvedt's (2015) *Water and Society* moves in the direction of making visible the relationship between water, society, history and politics but the changing perspectives addressed therein are not from a media studies perspective.

A scientific story of water (as a global narrative) cannot account for the particular experiences, enchantments and embodiments of water in highly mediated contexts wherein a 'watery sense of place' (Garde-Hansen, McEwen and Jones 2016), living with water, water as a living thing and water as a story thing (see, for example, www.biologyofstory.com) are integral to any national or cultural narrative. We cannot assume to understand a community's water stories (or even that a community knows it is a community because of its relationship with water) outside of the media genres, practices and texts pertinent for and to that setting. Many communities come into being through a mediated realization that water is so fundamental to their historical identity, or a new identity, as in a flood event or a migrant crossing of the sea, wherein members form collectives for action and support. This does not mean that communities created around a water issue are inclusive. For example, women and people of colour have a crucial and active part to play not only in recuperating their own stories of water but also in water governance and participatory politics wherein a watery (or dry) sense of place needs to be inclusive of marginalized communities and individuals.

Such a trajectory of restoration is in evidence in, for example, *The Political Ecology of Women, Water and Global Environmental Change* (2015) edited by Buechler and Hanson, and this is a starting point for considering these issues globally and more connectively. Yet this collection ranges widely from Brazil to Los Angeles, from Canada to Tajikistan with less attention to media. The specificity of the media infrastructures that have developed in those contexts and the use of media to creatively express stories of water are vital to explore. We need more research evidence for supporting new articulations of drought, flood, scarcity and water management through consideration of genre, the cultures of identities, such as gender and race, and the generational perspectives and learning from past water events. Thus, in a collective, connective and cultural memory sense, knowledge from one generation's mediatization of water ought to be transferred to or inherited by the next generations, even in the same subnational, regional or local setting.

Media and Water: Communication, Culture and Perception seeks, then, to champion a deeper understanding of 'water' by taking not only an arts and humanities approach but also one that focuses on research on media, cultural and communication discourses, forms and practices that currently and creatively enframe our reception of, experience within and calls to action around water issues. The historic and contemporary media representation of water, as both risk and resource, and water's entanglement with media practices, technologies and infrastructures, in a national media context (specifically in this book concerning the UK) is presented as connective with other contexts where water is undergoing similar *media treatment* (the Netherlands, Germany, France, the United States, Australia and Brazil, for example). Water is filtered and coded as 'good' or 'bad' while 'weather' and its extremes have been imagined in terms of war and conflict, and although snow and ice will not feature in this book, it is worth noting their construction in our imaginations as repositories or archives of freshwater.[32] While media are 'weatherised' (Twitter storms) and made liquid (immersive media and flow) so too a deeper appreciation of water as mediated (by formal and informal communicators) and mediatized (by official and unofficial outlets) is required.

It is timely to turn our attention to the mediation of water issues by a whole range of stakeholders with vested interests. At the heart of a media, culture and communication research approach to water are the following questions: where and

---

[32] See Endfield's 2012–13 Arts and Humanities Research Council funded *Snow Scenes: Exploring the Role of Place in Weather Memory*. Moreover, a typical representation of ice in popular media was noted during writing. *The Secret Life of Ice* (BBC4) was broadcast on 23 December 2018, with the synopsis on the author's television screen reading: 'Dr Gabrielle Walker looks deep within the ice crystal to try to discover how something so ephemeral has the power to sculpt landscapes, preserve the past and inform the future.'

who is the audience and how are they being related to, engaged with and enabled in creative responses? How are they being constructed, targeted, addressed and provided with material that informs, educates and entertains (to assume public service broadcasting values)? To what extent are they invited into a relational experience with the mediation of water that offers insights into the social and material aspects of water and media? How do the extant media templates shape and reshape the concepts of the water-user and the regional, national and global authorities, agencies and organizations that users are connected to? What does it mean to be positioned as a water stakeholder (usually rooted to physical locations) in these media templates and does this fully address the mobility of the water-user and the mobility of media? How are these relationships becoming increasingly socially mediated and connected online and revealing the connectivity of water across domains (physical, economic, political, spiritual, domestic, urban, public, private and personal)? All these questions address the most recent water research within anthropology (Wagner 2013), drawing on Mauss's (1990: 3) water as a 'total social fact' to Orlove and Caton's (2010: 403) 'waterworld[33] as the totality of connections that water may have in a given society'.

Therefore, a media, communication and cultural studies perspective is offered as attentive to textual and visual representation, audience and community, infrastructure and digital technology, new forms of social mediation as well as storytelling, remembrance and citizenship. This book offers a coherent, coordinated and evidence-based approach to understanding the history of mediating and communicating water in the UK, in particular, in terms of drought, floods and water management from a researcher who has worked across several nationally and internationally funded research projects with hydrologists, agronomists, ecologists, geographers, big data scientists, artists and ethnographers, as well as with flood-risk managers, water stakeholders in public and private sectors, civil society and in contexts as diverse as the UK, Brazil and Germany.

What is striking, working across all these diverse domains and sectors, is that each domain (academic or otherwise) and each sector (commercial or NGO) now use media, communication and culture as the key entry points for public and business-to-business engagement with water governance and for managing public perceptions, expectations and participation. Yet they do so, often with very little understanding of the mediated histories of water that have shaped publics, consumers, audiences, readers, viewers and users for decades and with even less appreciation of how these have used and are already using media and communication to reshape their own experiences and representations of water for new kinds of circulation. Thus, this book focuses on the representation of water in and through media as well as the audience reception of stories of water in a variety of formats and genres. More specifically, the

---

[33] Orlove and Caton (2010) also note the use of the term 'waterscape' since the mid-nineteenth century but used more since the work of Swyngedouw in his research on Spain (1999): 'He draws on political economy approaches within geography to examine the production of waterscapes, emphasizing the ideological dimensions of place in the construction of dams and canals and the creation of new administrative units based on watersheds' (Orlove and Caton 2010: 408).

book addresses the use of media by members of the public to recuperate and articulate hidden and marginal stories, producing new water discourse in a digital age.

In what follows, each section – Communication, Culture, Perception – begins with a chapter that offers a literature review pertaining to the structuring of that themed section and is then followed by two case study chapters that amplify and illuminate the key issues of that theme. Taken together the book offers a sustained and comprehensive exploration of the media representation, modes of address and communication frameworks that have constructed understandings of water in the UK context from 1947 to 2017, a period during which most countries have witnessed rapid media, communications and cultural developments. The book draws on a wide range of media forms from newspapers to digital, photography to radio, television to video, while more specifically focusing on a key area of expertise 'media and memory studies' in the form of empirical research findings from the many research projects that have informed this book.

Part One

Communication

1

Media templates for representing water

Introduction

Contact any environmental correspondent from a major newspaper to offer a piece on the scientific communication of water, and they will ask for something 'new' to say (new data, new findings, a new development) and a 'hook' (an anniversary, an event, a discovery, a memory or a 'water week'[1] perhaps) upon which to hang the story. It is not in the news media templates for water stories that we will find the exposition of diverse, slow-moving and complex relationships with water, for they, in themselves, create 'watersheds' – administrations of the stories of water that are acceptable to mainstream media. They have a long history of doing this that has been little researched. Thus, media templates expose the conceptual boundaries of media producers as Orlove and Caton (2010: 407) caution about 'watersheds', which may create regimes, practices and policies for managing water within nations and communities and across scales. Like media which are also shared, withheld, channelled, stored and deleted, 'water moves in many ways' and 'watersheds are not always the well bounded management units that water managers and others often assume them to be'.

There is, then, a growing importance ascribed to the role and impact of increasingly open media (particularly media images) in potentially shaping and influencing cultural understandings of complex geographical and scientific issues (Galaty 2010; O'Neill and Smith 2014).[2] On the one hand, images can be said to 'make and shape the world in visual and narrative terms' (Strüver 2007: 685). Yet, on the other hand, if we focus only on format we may miss the impact of sound, the written text and the affective experience which cannot be captured in visual terms. This suggests that a basic question is not being addressed here when thinking about communicating

---

[1] Hervé-Bazin lists the main international, regional, national and local 'water weeks' in the world (2014: 102).

[2] Consider the association of Greenpeace with the iconic Rainbow Warrior sailing the seas; WWF's poster campaigns against fossil fuels destroying oceans; the 2015 'Your Convenience Is Their Extinction' series by Christian Waters (advert designer on plastic pollution); protest and campaign imagery such as the thirty-three million views on YouTube of the sea turtle having a plastic straw removed from its nose filmed by Christine Figgener, marine biologist at Texas A&M University. One only needs to Google image search 'water conservation' to see many of the results are directed at or created by children, are dominated by the colour blue and usually feature an image of a blue planet, a running tap or a raindrop.

water. What are media? Meyrowitz (1993: 55–6) addressed this question directly and early on in 'media studies' when exploring the plethora of metaphors being used to describe different forms of media (metaphors I noted earlier as often hydro-symbolic) which he summarizes as '*media as conduits, media as languages, media as environments*'. With conduits expressing how media deliver content within contexts to receivers, this is the most common metaphor argues Meyrowitz; followed by media as languages (with unique grammar and aesthetics specific to their form) and media as environments, settings or contexts in themselves (1993: 58–60). So, while on the one hand media shape our understanding of water by acting as channels for delivering messages about water across various forms, one can also expect there to be a 'plasticity' to the medium, to borrow Meyrowitz's term, with many variables, manipulations and codes built into say a TV broadcast compared to a poster, a fictional film compared with a print news item. Finally, when media are considered environmental then we are looking at lives, emotions, remembrance, senses and thoughts in a relationship with different media discourses, forms and practices that play out in increasingly digitized and networked contexts.

This chapter addresses those key texts of academic literature from the studies of film,[3] broadcast media, journalism, participatory and online media that have addressed the infrastructures, representations and sociocultural discourses and images used to engage audiences in water-related 'events'. This media-focused literature is set within the context of a wider appreciation of climate and environment-related issues from an arts and humanities perspective[4] and does touch on social sciences approaches to media studies. While many of these texts can be applied to the mediation of water as a natural and managed resource, there is a handful of research literature (covered in later chapters) within the study of media, communication and culture that directly addresses the representation of water events (storm, drought, flood), water more specifically (rivers, seas, oceans) and water management (risk communication, citizen engagement, infrastructure and decision-making). Thus, mediating water is inevitably interdisciplinary, and undertaking research of media templates of water representation is likely to involve interdisciplinary methods. More recently, there is recognition that communication of water issues in a geographical, social and cultural context must take account of pre-existing lay knowledges in specific water cases (see, for example, Zeisley-Vralsted 2015; Coates 2013; Cusack 2007 on rivers and identity) and existing local participation in water management (for example, Tapsell 1997 on river restoration). This specificity is framed by overarching policies seeking co-production opportunities

---

[3] Cinema and water require a wholistic study of their own both in terms of histories of cinematic treatments of water, liquidity, storm and flood from cult cinema, sci-fi, disaster movies and pollution exposés (from around the world) to the use of film for documenting water and flood/drought issues.

[4] See, for example, 'Climate Stories' a Natural Environment Research Council funded project, which ran 'Three Days of Creating' at Dartington Hall, UK, in May 2018. See also the Hurricane Digital Memory Bank, http://hurricanearchive.org/, collecting and preserving the stories of Katrina and Rita at the Roy Rosenzweig Center for History and New Media (CHNM) at George Mason University and the University of New Orleans. The personification of 'water' by the Hollywood actor Penelope Cruz for Conservation International on YouTube was also part of series using celebrities to give nature 'a voice' during 2014 and clearly frames water's 'voice' in particular ways.

for dialoguing with communities on water-related issues[5] and deeper inclusion with media, communication and cultural studies. The latter have for a long time concerned themselves with race, ethnicity, gender, ethics and a politics of care and hope, as well as non-human rights and non-representational theories (see Buechler and Hanson 2015 on women, water and environmental change, for example).[6]

This chapter will argue that media, communication and cultural studies have a key role to play here and that this field of research cannot only be confined to issues of political or consumer communication and news reporting on the environment but must also take a lead (as conduit, language and environment) in defining the cultural and social principles and values of water as a cultural resource as well as natural capital. Put simply, natural resources are also cultural resources and water has cultural value and media values and not only in capitalist terms of reference. Professional communicators of water-related topics in a wide variety of sectors (such as engineering, private enterprise and environmental agencies) are experimenting with new ways of representing and communicating water. They often restrict and stage their water communications for the next 'water-related event' such as a flood-rich period, a drought-risk season, a water management policy change or even during a pandemic in which sewers become blocked by baby wipes and domestic water use increases significantly. Then they often forget to communicate with the new media content producers (formerly known as the audience) about everyday water issues in connective and relational ways. Therefore, we need to consider the 'art of the possible' in communicating water.[7]

In setting the scene of mass media templates, in what follows, one ought to highlight the underlying media and cultural infrastructures that have become so deeply intermingled with the natural resources and environment upon which those infrastructures depend. Such an intermingling will become more apparent in later chapters concerning what media archives reveal about broadcasting histories and the social mediation of the environment through digital media in the twenty-first century. While science communication, as an emergent area of research and practice within a social sciences perspective on media, will not be far from view, I believe that the arts and humanities offer both old and new approaches to lay expertise and personal voices noted as important and addressed later in the book, which are being enabled by social media and new communication platforms (see van House 2011; and Chapter 3). Often,

---

[5] See WWDR4 (2012) *Managing Water under Uncertainty and Risk: The United Nations World Water Development Programme*; UNESCO (2013) International Year of Water Co-operation; the open source UN Water (2013) Water factsheets and the UN Org (2013) International Decade for Action 'Water for Life' 2005–15. More locally, there were the Mayor of London assemblies on 'water' 2011–17 and the EU Water Framework Directive (2013). The latter makes a clear proposal for public participation based on the Aarhus Convention.

[6] In her 2016 Lecture 'Let Them Drown: The Violence of Othering in a Warming World' (London), Naomi Klein makes a case for re-appraising Edward Said's *Orientalism* (1978) as illuminating and clarifying 'the underlying causes of the global ecological crisis [...] that points to ways we might respond that are far more inclusive than current campaign models'. Said's defence of a right to return home is, says Klein, 'deeply relevant in our time of eroding coastlines, of nations disappearing beneath rising seas, of the coral reefs that sustain entire cultures being bleached white, of a balmy Arctic'.

[7] I am grateful to Jacqui Cotton of the UK Environment Agency for this insight.

these are in direct response to mass media templates of flood, drought and storm, for example. Overall, in drawing attention to the literature available concerning the mediation of extreme weather and natural disasters, we find water taking up a more nuanced position as part of wider contexts and cycles of its circulation (all of which are becoming mediatized or even smarter through digital sensors and citizen science). Thus, the sustainability (or not) of media representations of a 'watery' or 'dry sense of place', which becomes 'evented', remembered, forgotten and recommunicated by a variety of media forms is the basis of this first chapter.

In the context of mass mediation what makes these representations and templates memorable? I would argue that different water events play out in different media in specific ways within and across contexts, thus suggesting a need for a closer examination of the discrete histories, forms, discourses and practices of single media forms (that are the distinctiveness of TV, film, radio, newspapers, etc.) while at the same time recognizing their digital and 'cultural convergence' (see Jenkins 2006) in the twenty-first century. In the next chapter, I address media distinctiveness and interrelatedness by focusing on the historical representation of Flood and Tempest within the BBC's broadcast archives from the 1950s. For now, though, it is important to acknowledge the increasing intermingling of media forms across platforms, which demonstrates through media convergence an understanding of water narratives as both stories (fabula) and discourses (syuzhet)[8] and as potentially measurable in terms of their cultural value as lived-in experiences.

Why 'Media Studies' of water?

'Media Studies' (often maligned by 'the media' itself and those who work in media) is now a substantial field of research and teaching in many countries and in many research-led universities. The field has become more focused on environmental issues (see Chapman et al. 1997; Allan et al. 2000; Cox 2006; Hansen 2010; and Rödder et al. 2012); the sustainability of media, creative and cultural industries; and audiences as producers of eco-conscious content. While Rödder et al. (2012: 5) noted in *The Sciences' Media Connection* that it may be possible that science is orienting itself towards media not just for public view but also in 'the criteria of relevance for knowledge production', they do so in the knowledge that the emergence of fields such as 'environmental studies' may seem more relevant to the public. This may have an impact on the legitimacy of science but so too have the 'masses' had an impact on the legitimacy of both journalism and scientific communication, increasingly owning and using the means of production.

While initial interest in those early studies of media and environmental issues (in the late 1980s) was informed by political discourses on and critical analyses of the representation of nuclear energy (see Corner et al. 1990), there has been a steady growth in media-focused research literature that overlaps with geography (Schwartz 1996; Rose 2008; Jones and Garde-Hansen 2012), climate change communication

---

[8] An explanation of these terms can be found at https://en.wikipedia.org/wiki/Fabula_and_syuzhet.

(Nicholson-Cole 2005; Doyle 2007; Manzo 2010; Boycoff 2011; O'Neill 2013; Kunelius et al. 2016) and using media to research the environment (for example, Pink 2001, 2007, or Lester and Cottle 2009). More recently, nuanced subfields of media research have emerged such as the study of eco-cinema within Kääpä and Gustafsson (2013), memory cultures and the Anthropocene in Craps et al. (2018) and mass media communication of the Anthropocene in the recent research of Sklair (2017, 2018, 2019). It is often at the interface of science and politics that these texts engage with media (as the mediator of that interface, perhaps as a digital dashboard) but it is not only this interface that should concern how media researchers have engaged with water. More specifically, media studies researchers have addressed the representations of national water systems, such as in Yee-Lok Tam's (2012) case study of 'water imaginaries' in Chinese and Taiwanese television or in Hageman's (2009) Suzhou River in ecological cinema; or, have considered very specific national and historical accounts of floods, such as in Trümper and Nervela (2013) on mediated flood memory in Hamburg.[9]

Taken together, this convergence of media researchers on issues of water and environment suggests a reintegration of questions of media representations and media audiences with questions of cultural and environmental policy. That is, that the governance of arts and culture in terms of 'values' (economic and moral) actually concerns natural resources as much as artistic, creative and technical resources. Thus, this reorientation of the field of media, cultural and communication studies around questions of natural, cultural and political economies suggests a strengthening of the field for the benefit of social science and science communication. While most of the recent research falls within the realm of eco-media critique, using the methodologies of textual, ideological, narrative and genre analysis to address the representation of environmental issues and risks, there is also a burgeoning of research that concerns itself with impact, transformation and policy relevance. Some arts and humanities literature are deeply theoretical, philosophical and political about the intersection between a highly mediated and technologized social landscape and the natural landscape, with some emerging attention to the intersection between media infrastructures and natural resource (mis)management (as in Parks and Starosielski 2015).

Nevertheless, the vast body of media and communication research pertaining to environmental issues falls within the social sciences spectrum of media studies and tends to be concerned with science communication (Rödder et al. 2012), the public understanding of science in terms of expertise (Whatmore 2009), journalistic practice (Kunelius et al. 2016) and the role of digital devices in social science research (Ruppert et al. 2013). While important, these studies do tend to place 'media studies' as in 'the service of' scientific research, enabling the science communication aspects of projects to better understand how to reach mass audiences, market their findings or messages, and use social media (for example) to engage new publics. In fact, considering the fast pace of media development, such approaches can go out of date quite quickly, such

---

[9] See also Edy (1999) 'Journalistic Uses of Collective Memory'. How far researchers have taken the media representation of water and water politics seriously, as in the representation of the Water Initiative in the first Netflix series of *House of Cards* (2013–18) or considered that the television industry and market are built upon 'watercooler' moments, is yet to be seen.

that using phrases such as mass media (suggesting the *massive passives* maligned by interactive media companies) can already seem quaint.

Within media studies itself there has been a burgeoning of research that may not be *scientifically informed* but does address how 'the environment' is being defined, represented and constructed by a range of powerful stakeholders, from journalists to artists, from corporate social responsibility (CSR) messaging to fan activists. Some media studies researchers have been active in environmental critiques of policy and practice within media industries and wider culture. Media policy studies of the 'greenwashing' initiatives in professional contexts (see Cubitt 2005, 2009 and Kääpä 2014) and in the political economy research of digital media's impact on rare earths (see Cubitt 2009 and Reading 2014) or on indigenous cultural landscapes have been joined by studies of the use of smart media technologies through environmental sensors.[10] Aside from what Miller calls a 'still dominant cybertarian position' (2016: 23) that celebrates new media technologies' capacity to create 'friends' of and for the environment, there are those who recognize that massive digital media development will have a direct impact on the natural world[11] even if cultural geographers see media as increasingly 'atmospheric' and 'elemental' (see McCormack 2017).[12] Thus, at a macro level in seeking to more fully understand the impact of a rapidly developing globalized media and communications technology and infrastructure (in both platform power and deep-sea underwater cables), we have seen local critique of the cultural and creative industries impact on the environment. These industries may be new economic deliverers of prosperity, urban regeneration, smart technology for development and alleged health and well-being but they are some of the most precarious, non-unionized and exploitative industries that will require significant and joined-up policymaking across both environmental and cultural scales. Despite the fact many of these industries are often small and may present as more eco-conscious, independent-minded and pro-consumer than the large multinational conglomerates they seek to disrupt, it is clear they require massive and networked computing power to tap into global markets and supply chains.

Therefore, media studies scholars have not shied away from the necessary scepticism concerning the positivist and constructivist role of media representation, media production and media consumption on environmental issues, climate change and natural resources. They have in some key ways interrogated claims of sustainability, inclusion and ethics (such as in Maxwell and Miller's 2008 polemic 'Ecological Ethics and Media Technology'), risk communication through media (see Hicks et al. 2017), 'powerful environmentalisms' in celebrity studies (Brockington 2008) and global media production in key regions such as the Gold Coast, Australia (Goldsmith et al. 2010). These are

---

[10] See Balch (2014) 'New Technology Uses Social Media to Keep Track of Water Levels', *The Guardian* 17 July 2014. Available at: https://www.theguardian.com/sustainable-business/technology-social-media-water-levels-business; See also Sandover, Rebecca (2014) 'While Ministers Dither on Floods, Social Media Springs into Action'.

[11] As pastoral as they sound, data farms and cloud computing make a significant impact on the environment.

[12] The reader may wish to consider the significant connectivity of the BBC and Met Office for much of the twentieth century, and how in the early years of BBC management, the military knowledge of the naval and armed forces personnel who shaped BBC policy post-war, meant extreme weather and flood protocols were tightly contained in a command and control of information to the public.

just some of the many examples of how media studies as a field has expanded through sustained interdisciplinary research, and more recently through working directly with the sciences on natural environment issues. Yet these disconnected and discrete studies have not pervasively influenced the field in the way identity politics has, and not in the light of an emerging concept of 'hydrological globalization' (that focuses on waterscapes and flows, amphibious representations and reshaping territory through a watery sense of place).[13]

Media studies approaches to researching water are, then, not simply interesting textual, visual and audio-visual readings, approaches and methodologies to add in to the production of scientific methodologies and outputs to better engage audiences and citizens. Nor are they only offering an historical and cultural account of an environmental event or process through detailed media archive research. More than offering routes to better understanding how to reach the public, media studies offer strong interdisciplinarity to water projects with cultural values at stake. Clearly, this undersells how scientific studies of environmental risk are engaged in the invention of pasts, presents and futures; how they collaborate with public perceptions, aspirations and fears and are constructors of concepts of 'the public sphere', an at-risk 'community', notions of 'civic society' or 'citizenship'. It also oversells 'data' as concrete evidence and reveals the gap between cultural values and policymaking. For what are seemingly clean and observable 'facts' (for example, on river flows, from river gauges or of abstraction licences) can be also be considered 'stories', 'anecdotes' or sociocultural 'messaging' about and from river dwellers, water companies and farmers. Viewed in this way, 'water data' (be they from sensors pinged to mobile phone apps or songs sung at the river's edge) can make water visible and observable as social and cultural values.

As such, this requires a science-narrative integrative approach that operates at a range of scales: policy, professional as well as local practice, community levels and personal-social-cultural experiences. These have been historically mediated by national broadcasters, newspapers and public relations departments but now 'the public' has the capacity to broadcast and socially mediate their own stories, anecdotes and narratives with different degrees of freedom and articulation depending on the national and political context. Therefore, the move within the sciences towards a deeper public engagement (see Holliman and Jensen 2009), to appreciating the popular cultural and national representations of water (see Cusack 2007) and increasing public understanding of environmental risk (see Devitt and O'Neill 2016) may have tended towards the social sciences spectrum of the field (see Allan 2002), but more recently has ventured into new arts and humanities terrains (see Leeson 2014, 2018a, 2018b).

In fact, just as theory uses liquid metaphors,[14] noted in the Introduction to this book, so too, media researchers metaphorize the environment in their framings of

---

[13] Cymene Howe and Dominic Boyer presented 'On Flood and Ice: Hydrological Globalization and the Rise of Water' at the University of Warwick (22 February 2020). They produced a Cultures of Energy podcast 2015–19 in which they discussed the installation of the world's first memorial to a glacier fallen to climate change.

[14] Sarah Whatmore (2009: 588) states, 'those moments of ontological disturbance in which those things on which we rely as unexamined parts of the material fabric of everyday lives become molten and make their agential force felt', thus contributing to the metaphor in terms of everyday life as becoming liquid.

media technologies and paradigms of media analysis.[15] In her 2002 article 'Unnatural
Ecologies: The Metaphor of the Environment in Media Theory' Heise's opening subtitle is
'Alive in the Sea of Information', and she makes a point that speaks to the always already
entangling of the sciences and the arts noted earlier; 'that media theorists took over from
urban sociology, where "human ecology" had in its turn developed out of a translation
of categories from biological ecology' (2002: 149). Thus, imbricated within media theory
from the get-go was a latent ecological metaphorization of developing media technologies.
These pervaded the field and account for the many 'holistic' and interconnected
descriptions of a mediated environment with its channels, flows, immersions, saturations,
floods and droughts of news, images, information and messages.

As such, the human and the technological are weatherized into a relationship
of natural, disruptive but also agentic capacities. 'The really subversive element of
Ecology', says Evernden,

> rests not on any of its more sophisticated concepts, but on its most basic premise:
> inter-relatedness. But the genuinely radical nature of that proposition is not
> generally perceived, even, I think, by ecologists. To the western mind, *inter-
> related* implies a causal connection […] but what is actually involved is a genuine
> *intermingling* of parts of the ecosystem. There are no discrete entities.
>
> (1996: 93, emphases in original)

In line with that premise, media theorists and practitioners have offered a radical
trans-mediality within and across media forms.

What is interesting in Heise's account above, and for this book as a whole which
seeks to intermingle but not dilute the disciplines, is the idea of return and bending
back, a form of resilience in our thinking (perhaps even a return to strong disciplines
in the face of interdisciplinary challenges) that 'allow[s] one to bend the metaphor
back to its literal context, and to investigate the interplay of technology and nature in a
more broadly understood spatial ecology that encompasses both material and virtual
habitats' (Heise 2002: 149). An example of such bending back through a media studies
approach to water (wherein material and virtual habitats intermingle in an analysis)
would be to return to the production of print-copy magazines and newspapers. This
is to reconsider the role of water in the pulp and paper industry by the end of the
twentieth century in the West, for example, or to consider how this industry became the
'single largest consumer of water used in industrial activities in the wealth democracies
of the OECD and the third largest greenhouse gas emitter, after the chemical and
steel industries' (OECD 2001: 128). To rethink media histories, media nostalgia for
old forms and new markets for traditional forms of media production, distribution
and readership through water history allows media researchers to understand
environmental issues not as simply mediated but as a problem for media industries
themselves. A recognition of the intermingling and what we were wrong about.

---

[15] In *The Rise of the Network Society: The Information Age* Castells argued that flows dominate life:
'our society is constructed around flows', he writes, 'flows of capital, flows of information, flows of
technology, flows of organisational interactions, flows of images, sounds, and symbols' (2000: 442).

This is particularly the case when we consider the growth of print news and magazine industries in China in the twenty-first century and the increased demand for and impact upon water by energy industries (such as in hydro-electrics and fracking) to service the exponential growth in paperless media and technology (tablets, phones, scanners, cameras, laptops and screens). Therefore, if the technologically mediated and converged system of TV, film, news, radio and social media is now the 'new' *water we swim in*, such that it is our invisible life force without which we cannot live an economic, cultural, social and family life, then how can we address this critically as more than just a metaphor? How can we bend back upon our current understanding of media, which seems as vital to us as water itself, with deeper and historical attention to the interrelatedness of media, climate, weather, nature and environment? One starting point is to consider the construction of climate in experience, memory, emotion and story. As Hulme et al. argue: 'For people living in particular places and particular cultures, climate is constructed as a function of their experiences and memories of past weather events, and what is socially learned from previous generations. These climates may often be reified through paintings or photographs of physical markers, such as a flood, drought or a rare snowfall' (Hulme et al. 2009: 198). Thus, the intermingling and entanglement of media and water become highly charged in terms of policy, management and public engagement, particularly in terms of greening (or blue-ing) media industries.

Water as a media event

An entry point to mass media templates of water issues concerns 'media events' covered in more depth in the next chapter. It is a concept developed by Dayan and Katz in 1992, who define media events to include the interruption of the routine, the monopolistic takeover of broadcasting channels and the live transmission of the event through media channels (Dayan and Katz 1992: 5,10), differentiating between 'a great news story' and 'a great ceremonial event' explaining that, for example, the assassination of President Kennedy was the former whereas his funeral was the latter (Dayan and Katz 1992: 9). There is a significant section of literature in media and communication studies (such as Mitu and Poulakidakos 2016) on theorizing and analysing 'media events' themselves as a breadth of discrete categories of the ceremonial; such as political (elections), national (royal weddings) and sporting (Olympics) events (see Dayan and Katz 1992), to name but a few. There are also disruptive events that have evolved Dayan and Katz's 'Contests, Conquests, and Coronations' (Dayan and Katz 1992: 1) to include 'Disaster, Terror, and War' (Katz and Liebes 2007: 157).

The evolution of the concept into a media 'eventscape' is enabled by the increasingly connected global media infrastructures and the live reporting of the public that reshape our understanding of media events today (see Couldry et al. 2010: 8).[16] If the main difference between media events and disruptive mediated events is the

---

[16] Indeed, Katz admitted that there has been a rise in the broadcasting of disruptive events which has subsequently led them to become included in the genre of media events, be it in a separate category (Katz and Liebes 2007: 163).

pre-planning of the former then this suggests that only the former has a mass media template that is oven-ready for broadcasters and the public (see Bacallao-Pino 2016). Yet clearly personal, professional, collective and cultural memory has a role to play at the interface between the event and the audience, particularly if this is metaphorized as a 'media scape' or more colloquially as a 'media circus'. Hoskins (2018) addressed this in terms of a 'global memory' of major news events, catastrophes and disasters as witnesses not dissimilar to Dayan and Katz's 'electronic monuments' (1992: 211) or Bacallao-Pino's (2016) transmedia events that create repertoires of collective action.[17]

The mediated 'water events' I propose here and develop throughout the book have had less attention from media studies scholars. While climate-related events bear some similarities to 'media events' such as being 'phatic' (see Dayan and Katz 1992) there are also important differences that concern the emergence of new techniques for recording and remembering climate and weather,[18] and the integration of science communication into an ongoing narrative representation of water's eventhood.[19] On the one hand, one can argue that 'water events' (i.e. drought, flood, tsunami, hurricane) which occur within an increasingly globally mediated ecology are articulated as 'certain situated, thickened, centering performances of mediated communication that are focused on a specific thematic core, across different media products and reach a wide and diverse multiplicity of audiences and participants' (Couldry et al. 2010: 12). On the other hand, the everyday mediation of water is rarely neither a news event nor a centring performance on a global stage; and it is these kinds of small and everyday mediated stories of water that I focus on in Chapter 3.

Therefore, and as I will show in this book, it is in and through the emergence of new kinds of media templates for representing the everyday performance of water use and management that science, environmental and policy communicators could seek an impact. Moreover, this filtering of media for impact should not ignore the smaller spheres of water story-ing that occur in the everyday encounters between members of the public and those who manage water and water issues. While experts' use of media may suggest a 'broadcast' approach of centring, thickening and situated-ness, they are being increasingly compelled to mediatize water through a post-broadcast approach of diffuse, networked and co-developed storytelling. Science and environmental

---

[17] One might also consider the social media conversations between 'media events' as opportunities for users to engage in creating contagious, viral or eruptive 'events' that snowball into 'media events'. I am thinking here of Twitter storms (covered later) that seemingly emerge out of egregious, or even, misunderstood tweets. It seems the Twittersphere, as it is so often called, is always already waiting to be 'evented' so to speak.

[18] Douglas Kellner categorizes media events as 'media spectacles', a term he believes is more suitable in the age of mass media and expanding channels of communication around the world (Kellner 2010: 76–7). I return to the 'spectacular' in Chapter 4.

[19] Consider, for example, storm-naming. The Met Office in the UK addresses storm-naming from a media and memorability perspective. If the storm is going to be disruptive enough, its impact and the likelihood of wind, rainfall, snow, ice to be significant, then it is more likely to be named. Public perception is high if the storm is named and when the name matters. It is noteworthy that female-named hurricanes are deadlier than male-named hurricanes (see Jung et al. 2014). For example, Storm Doris gained a good deal of traction on social media.

communicators on water issues now realize, as did Napoli in his book *Audience Evolution*, that the audience is becoming more independent and controlling 'over when, how, and where they consume media' (Napoli 2011: 1).

Thus, it is in the period between the newsworthy 'water events' of flood, storm and drought that researchers need to work, wherein the public's relationship to water is established and maintained or disappears altogether. It is not that the public is not interested in water, or loses interest, but rather those communicating water may often only en-frame the topic in newsworthy templates of 'eventhood', marketing terms of customer relations or public communication for behavioural change and ignore the everyday media templates, produced through longer duration narratives. A good example here is the BBC radio drama *The Archers*, which has been broadcast from 1950 to the present day,[20] and has incorporated water stories not only into its scripts but more recently into its social media communication, and character development on the programme's webpages, which have in turn been re-mediated by non-governmental organizations such as the National Flood Forum. Thus, if the science domain were to recognize that media and water have been bedfellows for quite some time, even at the level of the early days of developing broadcast infrastructures, as Caughie notes in his history of television[21] for example, then the integrative approach of science and narrative would not be so challenging.

Water is mediated, produced through increasingly converged media technologies and considering the 'spreadability' (Jenkins et al. 2013) of media, media production and media consumption, we find water mediated by both water industries and environmental agencies in increasingly creative ways. Media powerfully shape meanings, experiences and audience memories, but they are also tools used by audiences to create their own meanings, experiences and memories. Citizens make decisions based upon their interpretations of media stories and increasingly use media to tell their own stories. This does not mean that audiences are necessarily directly injected with messages (see Katz and Lazarsfeld 1955 'Hypodermic Needle Model'), or coerced into doing what media tells them, they do *and* they do not resist the very mechanisms that seek to turn them into an audience in the first place.[22] Rather, in mediating water events we should first acknowledge that viewers, readers, users and consumers are highly media literate within their own memorable and emotionalized contexts or spheres of mediated cultural production and consumption. Apart from the odd journal article that may draw upon media evidence (newspapers, film, television, social media) often from another disciplinary perspective, there is very little from a more specific media and communication studies perspective that focuses on spheres of mediated water issues, stories and challenges.

---

[20] See BBC Radio 4 *The Archers* blog 'The Brave New World of Adam Macy', 11 April 2016. Available at: http://www.bbc.co.uk/blogs/thearchers/entries/36bc4e11-4ab7-4d5d-b691-6326aa0f797b.

[21] British television began life following 'a pattern established by the late Victorians and the Edwardians for the administration of other national utilities like water, gas, and electricity' (Caughie 2000: 26).

[22] Early on Bernard Cohen stated that mass media 'may not be successful in telling people what to think, but it is stunningly successful in telling its readers what to think about' (1963: 13). This is called 'agenda-setting' and framing (see McCombs and Shaw 1972).

Mostly, scholarship in the field has focused on climate change and mediated emotions[23] such as Wang et al. (2018) on emotions predicting policy or Chapman et al. (2017) on reassessing emotion in climate change communication. When texts do cite 'media' or 'mass media', they tend to restrict their view point to 'the media' meaning 'the press', broadcast news and public relations and dismiss these as engaging in the tabloidization of water events (only interested in news values of drama, tragedy and celebrity campaigning), trivializing the environment or on the side of PR and marketing strategies. What is not addressed is the role of mediation or mediatization of water as culture, remembrances of water as well as future scenarios: not simply in representations of flood and drought as, for example, in *Extreme Weather and Global Media* (2015) edited by Leyda and Negra, but in everyday water cultures of Allon and Sofoulis (2006) or in remembering Hurricane Katrina in Robinson (2009). While texts such as *Reconnecting People and Water: Public Engagement and Sustainable Urban Water Management* (Sharp 2017) are important to understanding this domain, they do tend to miss the important role of media in that connection and reconnection from the get-go. Framing media in the management of water as misreading, misrepresenting or manipulated by vested interests is to assume that water is not always already highly mediatized and story-fied. This can mean that water-stressed communities are underserved by media and underserved by cultural organizations who specialize in climate adaptation strategies. Yet there are many fronts of change: new roles for the citizen (from modern to ecological), the backlash to nostalgic nationhood and colonialist attitudes to the environment, the rise of digital participation, the empowering of young people through social networks and movements, the bringing of less heard voices to the water governance table (addressing flows of power and privilege) and the daylighting of hidden waterways in cities. All of these involve narrative, storytelling and media, and individuals and communities are demanding inclusion.

Can media and culture be built into water management and policymaking? Pasotti seems to think so and describes the new kinds of media templates for water issues produced in Bogotá, Columbia, in 1997 through 'symbolic policies' and a 'new role for citizens' where the 'mitigation of risk was not borne by the government alone but shared by all citizens'

> because of a tunnel malfunction, the city faced substantial water shortages, which called for investment in a new water reservoir. [...] In several television appearances the mayor explained the situation and provided consumption-saving tips. He even broadcast from the shower giving the audience water saving advice, in a supremely pedagogical exercise.
>
> (Pasotti 2013: 49)

---

[23] Cultural critic Naomi Klein is vocal on this: 'The point is, today everyone can see that the system is deeply unjust and careening out of control. Unfettered greed has trashed the global economy. And we are trashing the natural world. We are overfishing our oceans, polluting our water with fracking and deepwater drilling, turning to the dirtiest forms of energy on the planet, like the Alberta tar sands. The atmosphere can't absorb the amount of carbon we are putting into it, creating dangerous warming. The new normal is serial disasters: economic and ecological' (Klein 2012: 45).

Can water issues be built into media, communication and cultural research? Many argue this is a necessity considering the eco-challenge that media technologies and industries themselves pose. Media studies is beginning to and will need to fully address:

> 1) The environmental burdens of energy generation and consumption throughout the medium's life cycle, from production to consumption and disposal, including transportation throughout its cycle; 2) a medium's chemical and heavy metal composition; 3) prior inputs from the earth (extracted via mining, drilling, logging, etc.) – the source function of the eco-system; and 4) subsequent outputs into the earth (deposits into air, land, and water [...]). The ecological dimension of media technology points to ethical questions that the field must not shy away from. The most challenging will be how much communication and entertainment media is enough to attain a system that serves everyone on the planet fairly without contributing to 'ecological suicide'.
>
> (Maxwell and Miller 2008: 347)

If media technologies 'carry both promise and peril', they are 'tolerant and good' as well as 'pose hazards' then how can 'eco-ethics' (Maxwell and Miller 2008: 347) offer an intervention into the field of media studies if the history of analogue media has ignored the superworkers of media (the pre-human work of oil, gas and water) that enable media production and cultural work? Particularly, as green media studies largely assumes terrestrial dimensions and fails to account for the long history of water use by media industries. The next chapter begins the process of rethinking the history of British broadcasting through water history.

Conclusion: Mediating through water

In *Thinking with Water* (2013) edited by Chen et al., they position traditional storytelling (privileging 'traditional wisdom', indigeneity, orality and folklore) as a 'powerful way to bring community together' in stark contrast to what they call the local or international media 'manipulation of resource management discourses' (2013: 9). This assumes that media's storying of water is a tool always in the hands of powerful organizations that consumers are not sovereign and civic-minded and that audiences are media illiterate. Media (in many domains) and in many forms have a long history of asking incisive questions about assumptions concerning how knowledges of water are shared, and how water creates communities (of action). Indigenous cultures and traditional wisdom are often recirculated by media and new kinds of visuals, aesthetics, performance and culture that play a key role in environmental adaptation (see McCumber 2017 on drought and shifting aesthetics in Santa Barbara, California; and, Brown 2017 on the variable visibility of water stress and water supply in four US cities as a visual perception of 'water, water everywhere').

Therefore, to place media in opposition to 'alternative ways of "story-ing" water and mapping waters' is to disavow the role of a variety of media discourses, forms and practices in shaping spheres of living with and without water, and the significant impact of say 'spectacular television', for example, which I will cover in Chapter 4. To claim that only authentic and unmediated ways of communicating 'can give voice to inclusive and evolving vocabularies of watery place, thereby transforming collective ways of thinking' (Chen et al. 2013: 9), misses both the history of broadcasting (which I will address in the following chapter) and new forms of representing water issues (focused on in Chapters 3, 5 and 9) and the contemporary possibilities for socially mediating collectives of water action (see Chapters 7 and 8).

Chen et al. (2013) are correct to draw attention to how urban infrastructures design water stories through narratives that either forget water (through canalization, for example, which then leads to stories of lost or missing water) or remember water through new stories of responsibility and water scarcity in a changing climate:

> Different ways of situating water may serve to acknowledge or deny our participation in and our obligations to communities enabled by these same waters. For example, many urbanites forget (or little recognize) how the health of a city and its inhabitants is premised upon the well-being of the surrounding waters [...]. Indeed, the way we choose to build our cities can severely limit our understanding of water and may even encourage its forgetting.
>
> (Chen et al. 2013: 9)

Remembering water (which I will deal with in more depth in Chapters 6 and 8) from a media and memory studies perspective is defined by Chen et al. (2013) as being 'responsible to water', as responsive to 'its articulations of kinship', to 'deliberately iterate and reiterate' the 'diverse situations of water'. This assumes a fairly straightforward relationship between storying water, creating 'inclusive vocabularies of watery place', and inclusivity itself; between being *responsibilized* towards water by discourse and feeling and acting upon that responsibility. It assumes that story leads to action and that water itself can be transformed into a tool of cultural policymaking. While here it is important to draw attention to the power of story-ing (later analysed in Chapter 8 on riparian media and culture), it also clearly explores the limitations of those iterative and reiterative practices if they do not impact decision-making, water governance or environment policy (which I cover in more detail in Chapter 7), as well as consider 'whose' policy is being privileged. By policy, I do not mean only environmental or economic policies, where conversations about urban water, for example, are likely to arise every day; but, rather cultural policy, where the cultural value of water is often never mentioned at all, and which I cover in the final three chapters of this book.[24]

There is another aspect that remains largely invisible: the media production processes (technical, physical and relational) as well as the agents, actors and stakeholders who are all playing (sometimes co-productive) roles in shaping and influencing the cultural concepts and relationships with water. As noted in the Introduction to this book, there is a missing set of literature from the arts and humanities research community that

---

[24] Take, for example, Richard Florida's (2004) creative cities thesis, which has influenced many national cultural policies around the world, through policy transfer, gentrification practices and globalization.

more deeply connects with scientific and social sciences approaches to water from a media and communications perspective.[25] New research on water ought to consider the practices and representations of cultural and creative actors in industries as diverse as theatre, television, film, radio, journalism, popular music, art, design and more recently social media.[26] These actors – or media, cultural, creative workers – who are embedded in public, commercial and entrepreneurial organizations, are engaging in techniques and practices that emotionally and memorably connect people on environmental issues. As part of creative cities, regeneration projects, well-being initiatives, street performances and social enterprise development, we find new kinds of a watery sense of place. They have the power to brand a city as 'green', 'blue', 'radical', 'technological' or 'heritage', ensuring that audiences, consumers, readers and users are primed to understand the connections being made and remade between culture and water.

---

[25] One key exception is *Water Communication: Analysis of Strategies and Campaigns from the Water Sector*, edited Hervé-Bazin (2014) more from a 'communications' paradigm.

[26] There is no space in this book to cover many of the cinematic examples, but it is important to note that audiences consume water messages in a rich media ecology. The documentary *Blue Gold: World Water Wars* (2008 dir. Bozzo) addresses private greed and public rights, and alongside *Flow: For the Love of Water* (2008 dir. Salina) and *Tapped* (2009 dir. Soechtig and Lindsay) all offer post-financial crash accounts of the human and the natural world at the mercy of corporate and global industries.

2

Deluge and tempest in the BBC archives

Introduction: Framing big floods

Media content on big floods of the twenty-first century now circulates widely. Flood events in one part of the world filter into public perceptions in another, so it may be difficult to make the case that *national* reporting on flooding sensitizes only that nation's public to climate changes any more or less than other reports from places to which a public has no specific connection. Nor can we be sure that framings of big floods have the intended impact on the target audience in ways that connect those audiences across time and space to environmental issues that are becoming of global concern. As a BBC *Woman's Hour* radio broadcast of 3 December 1954 wearily lamented:

> I think that it is terribly difficult for us to get an overall view of this sort of thing. Tempests, floods and storms, terrifying manifestations of natural forces though they be, only really come alive when they touch us personally – when the roof has been ripped off our house, or we see our furniture vanishing under a swirling, dirty mass of water. Oh! It's true that we do make some kind of response to the unemotional voice of the news announcer, or the page-wide headlines in the press. 'How sad!' we say, 'How dreadful for those poor people!' 'I wouldn't like to be a sailor in this kind of weather and so on.' Quite possibly we make a contribution to some flood disaster fund. And that is that.[1]

Thus, it is not a recent concern that the media framing of big floods struggles to deeply engage and connect audiences in a sustained way, even if the audience resides in a flood-prone region or nation. There has, in fact, been an ongoing conversation about how to represent flooding through media in the UK since the 1950s (as the broadcast script above reveals). Nor is it new – or specific to our digital age – to expect flood media to circulate well beyond national boundaries for global audiences because the media infrastructures of global communications now facilitate this. This makes two limiting assumptions: that analogue media did not circulate before the internet and that localized and deeply rooted media is out of date in a global media market.

---

[1] BBC *Woman's Hour* broadcast script entitled 'Behind the Headlines'. Producer: Peggy Barker. Transmission: Friday, 3 December 1954, 2–3 p.m. The *Light Programme* script files. BBC Written Archives, Caversham.

In fact, the BBC Written Archives hold substantive evidence that, during the East Coast Storm Surge in the winter of 1953,[2] deep connections were forged across nations that shared the same experience of death and destruction caused by natural disaster. For instance, both the Netherlands and the UK exchanged donations of clothes, food, money and household items, and rescue services worked between the two countries. As one *Home Service* broadcast on 3 February 1953 shows, reported from The Hague by Lionel Fleming:

> The people of the Netherlands are just beginning, as it were, to draw breath again after the calamity that fell on them so suddenly during the weekend. Even now it's impossible to tell the full scale of the disaster. Nobody can tell you what the final figure of dead is likely to be [...]. It would be fair to say, I think, that the people themselves took action. At any rate the rescue work owed a very great deal to purely volunteer effort. [...] even radio enthusiasts played their part and gave valuable information about the state of affairs in their own district. And, of course, one mustn't forget the help which came from outside. Britain's readiness to help, and the part already played by the RAF helicopters, have been very sorely welcomed here.[3]

What emerges from the broadcast media archive is that floods become media events through a variety of diverse media, cultural and communication practices and are produced by and in that location for a variety of proximate and distant audiences, which can produce both nationalism and empathy.[4] More interestingly, in the current context of social media reporting, it is noted in the radio broadcast above that citizens were called to action and that through radio enthusiasts media itself took on an amateur reporting role at the grassroots level.

Therefore, recent scholarly accounts of the media framing of big floods without reference to media history should not entirely shape our understanding of the mediatization of water events. In *Flood of Images: Media, Memory and Hurricane Katrina* (2015: 9) Bernie Cook is deeply critical of the 2005 American television news culture surrounding the flood of images that seemed to racialize the victims of Katrina as 'violent, dangerous and depraved'. While this position accords with the notion that media (albeit global and internationally connected) are local, ideological and contextually specific in their representations of flood and storm, it misses the audience's longer memories of historical mediations of flooding in their particular

---

[2] 31 January 1953 to 1 February 1953.

[3] See the BBC *News Talks Feature* entitled 'Floods' broadcast script Producer: Transmission: 3 February 1953. Producer Bill Northwood. Narrated by John Snagge with reports from Dr R. A. Schofield, Sylvia Gray, Valentine Selsey, Philip Donnellan, Ivor Jones, James Bell and Lionel Fleming. It is worth noting that Philip Donnellan would go on to be a highly successful documentary maker. The *Home Service* script files, BBC Written Archives, Caversham.

[4] The BBC Written Archives show in great detail the cities and countries who came to the assistance of the Netherlands: UK (Coventry, Devon, Glasgow and S. England), Norway, Italy, France, Sweden, Germany, America, Argentina, Switzerland, Indonesia, Belgium, Canada, Israel, Egypt, Ireland, Malta, Russia, Iceland, India, Turkey, Portugal, Peru, Philippines, S. Africa, Australia and Finland, for example.

place alongside a much wider repertoire of mediated accounts of deluge and tempest from a range of genres, places and peoples. It reads flood representation through political and social history only rather than alongside an environmental history of flood events and samples media forms, such as television (as in Cook's work cited above) from a repertoire or ecology of 'weatherized' media to which audiences have access in the twenty-first century. Such readings may be helpful in considering racism in mediated flood events but less helpful in understanding the relationship between flood risk and race at the community level, and the role of different forms of media in that relationship.

Furthermore, I would argue that audiences can be sensitized to discourses on flooding as much by the representation of deluge and freezing in the blockbuster film *The Day after Tomorrow* (2004, dir. Emmerich)[5] as by a local news report on massive snow melt, as much by your cultural tastes, age, home ownership status as your willingness to heed *The Guardian* newspaper's environmental reporting. Media literacy, accessibility, cultural values and trust in journalism all play a role in the complexity of media and communication messaging and reception. Moreover, media producers know this and their production strategies of storm and flood have become interpolated into and borrow from other media genres, from Hollywood film to television drama.

Cook makes a key point concerning media templates and Hurricane Katrina in the context of US television networks, '[l]ocation matters to the production and uses of memory as to the production and uses of media' (2015: xiii). In her interview with a journalist she learns of the 'hurricane playbook'. That is, those 'predetermined moves' that feature the following: 'the storm track, interviews experts […], checks in on preparations, and interviews a few people' with 'the reporter out in the storm […] to be the most important visual', which is 'the "money shot" that confirms the danger posed by the storm, the live-ness of the coverage […], and the mastery of the television news apparatus to bring this managed danger to audiences' (Cook 2015: 9). While this suggests templates (as noted in Chapter 1), it also implies television's own amnesia playing a role in flood memory by continually broadcasting the unprecedented weather event as new without referencing its own archive. This mastery is no less evident in the 1953 BBC radio broadcasts with reporters stationed alongside rivers and coasts in England and Holland, and reporters reporting from the air.

However, rather than place media and flooding in distinctive and disconnected positions, pitted against one another, this chapter argues that media histories have been intertwined with water histories for a long time, at least in a UK context. Water histories have been materialized and formalized in media archives (print and broadcast) and revealing these histories as forms of cultural memory can offer water scientists, social, cultural and media historians new ways for understanding and articulating living with extreme weather in a changing climate, which extends back from and beyond a climate change discourse. In later chapters, I will draw upon social media for mapping flood

---

[5] In *Old and New Media after Katrina*, Diane Negra (2010: 13) makes the point that a rise in Christian apocalypse fiction, rapture fiction, 'contemporary necropolitics' and computer games of apocalypse, shaped audience responses to Katrina, thus suggestive of a wider media ecology to floods.

memory but for now it is key to the mixed-media approach to frame the 1947 and 1953 deluge and storm as part of a growing *memo-techno-ecology* (see Garde-Hansen et al. 2016) of remembering and forgetting water crises over longer time periods and in their specific media contexts.

Daylighting[6] broadcast archives

Firstly, one needs to make the case for broadcast archives, those (often) analogue repositories of knowledge, experience, representation and narratives of the twentieth century that predate the climate change discourse and can be tricky to resurface to explore how past flood events have shaped national identity, temporality and media infrastructures. I have sought to 'daylight' the production-based evidence of storm and flood in British media history through drawing together print media, broadcast media and public engagement with images of the Winter Floods of 1947, and broadcast material from radio and television of the 1953 East Coast Storm surge. Both these events were remembered and remediated in the Summer Floods of 2007 (which took journalists by surprise)[7] and the East Coast Storm surge of December 2013 (which did not take journalists by surprise).[8]

Thus, researching media archives draws upon research from the UK ESRC funded *Sustainable Flood Memories* (2010–13) project,[9] the findings of which led to several publications (Krause et al. 2012; McEwen et al. 2012a, 2012b; Garde-Hansen et al. 2016; Garde-Hansen et al. 2017; and McEwen et al. 2016). This project was concerned with learning to live with water, making room for water in one's memory and imagination and being prepared for water scarcity. It chimed with wider understandings of risk and resilience, such as in Cutter et al. (2008) on how disaster can 'be moderated by the absorptive capacity of the community [...] leading to a high degree of recovery' or that a 'community may exercise its adaptive resilience through improvisation and learning', the latter defined by Adger as 'the diversity of adaptations, and the promotion of strong

---

6 Wild et al. in 'Deculverting: Reviewing the Evidence on the "Daylighting" and Restoration of Culverted Rivers' in *Water and Environment Journal* who define this as 'opening up buried watercourses and restoring them to more natural conditions' (2011: 412). I use the term here metaphorically to daylight cultural memories through opening up, re-routing and bringing to the surface media archives and cultural content that also adds value to communities. See also http://www.daylightingrivers.com/.

7 The *Gloucestershire Echo* of 21 July 2008 had a double-page spread mostly taken up with images of submerged cars and houses with the headline: 'Deluge brings county to a halt'.

8 News producers will be *copy ready* for anniversaries, setting the agenda in anticipation of a storm event during a flood-rich period based on mediated memories. The storm surge of December 2013 framed the event as the most serious for sixty years and repeatedly referred to the 1953 Storm. See BBC News Report 'Tidal surge hits east UK coastal towns after storm', 6 December 2013. Available at: https://www.bbc.co.uk/news/uk-25253080. Accessed 6 December 2018.

9 The *Sustainable Flood Memories* project built upon the AHRC funded Living Flood Histories Network (2010) and both offered an innovative and interdisciplinary approach to memory and histories of environmental change. We studied four floodplain settings in the lower Severn Valley, UK, and integrated the team's expertise in flood-risk management, cultural geography, media and memory, social anthropology and oral history.

local social cohesion and mechanisms for collective action' (Adger et al. 2005: 1038). Cutter et al. (2008) are important to cite in the context of media and memory because they point to institutions (and one might add, archives) as playing a particularly important role:

> Social learning occurs when beneficial impromptu actions are formalized into institutional policy for handling future events and is particularly important because individual memory is subject to decay over time. Manifestations of social learning include policy making and pre-event preparedness improvements. When improvisation and social learning take place, they directly alter the inherent resilience for the next event.
>
> (Cutter et al. 2008: 598)

It is in this wider programme of connecting communities around environmental issues, exploring watery scapes, understanding episodic extremes and casual flooding from the perspective of cultural memory and heritage that British media archives emerge for social resilience.

The two case studies below are based on three positions. Firstly, media stories of flooding are as inheritable as family stories, and intertwined; for, it is increasingly the case that flood experiences are mediated by the audience as they become media producers. Secondly, British media attention to flooding and water more generally has a much longer and deeper history than the current twenty-first-century online accessibility of media stories and images give credit for. It is more the case, which neither broadcasters (public or commercial who see these stories as niche) nor researchers (who forget they are there) have demanded much access to the media archive (which is not always organized around the flood event but by the provenance of the producer), and so they are seen as out of date or out of sync. Thirdly, mediated memories of floods are widely remediated in specific milieu, creating bubbles of experience and knowledge, which without closer attention to media archives and their contents can lead water researchers to assume that media directly reflect public opinion and vice versa at that given moment in time.

Case study 1: Sustainable flood memories of 1947

The 1947 Winter Flood was remediated in 2007 as a cultural and collective memory reactivated in the minds of contemporary audiences of flood media through a resilience narrative that plugged audiences back in to a longer flood media narrative.[10] As we noted recently, the public

---

[10] On 25 July 2007 in the aftermath of the Summer Floods, *The Guardian* newspaper journalist Martin Wainright offers an excellent synopsis of the memorialization of flooding in British culture. He argues that 'the country has few equals in putting up memorials to great soakings of the past. Everywhere from York to Gloucester via London, notched poles mark the riversides, engraved with historic high-water levels. Prominent on them all is the date 1947, the benchmark year in living memory for every subsequent flood.'

were being re-mediatized as connected across time and space and in terms of collective memories that often incorporated 'the Blitz spirit'.[11] Many of our older interviewees referenced the Second World War, which only ended two years prior to the 1947 floods, as an important marker of British resilience to disaster. This historical connectivity interwove 'living with water' as both a continuous activity and an extension of a wartime morale that must be quietly maintained.

(Garde-Hansen et al. 2016: 62)

The risk and resilience narrative of British national identity plays a strong part in the association of weather with war (see Massumi 2011 and Furedi 2007[12]) not only in the minds of the public today influenced by a resurgence of globally marketed nostalgia (i.e. *Keep Calm and Carry On*) but also in a *business as usual* discourse.

Media producers play an important role here as noted in our interview with a BBC journalist on his reflections of interviewing a flood victim:

I filmed him going through his house and the water had gone away so it was all full of mud and goodness knows what and we were pulling out drawers and the water would fall out [...] and he said, 'Oh God, that was my army ... I was given in the Second World War' and you know, first thing you've got great television, great emotion and he just laughed and went 'so what can I do?' He said 'I can't get upset about these things, it's happened, if you got upset about it you'd screw yourself up' and I think it's important to show that resilience as well, against a guy who fought in the Second World War for God's sake so a bit of flooding isn't going to help him but you had that tangible human story that he was losing prize possessions and he was saying at least I'm still here, I'm alright, and those stories are really strong. When it comes to flooding I think those are the stories people want to hear.

(Interview with BBC Media Producer, 15 May 2013)

While floodplain residents were a source of living and transmittable memories of resilience, who had access to long histories of flood materialization, they were bolstered by the monumentalizing of flood memory of 1947 as inscribed on walls in the region, such as the ones below (see Figure 2.1). Such images were widely remediated through connecting photos of 1947 flooding with photos of 2007 flooding and this materialization was sometimes referred to by journalists who would remediate images of the town's abbey surrounded by water to iconic effect.

As in the case of the 'hurricane playbook' noted above, stories of big floods and community resilience are produced from templates as Hoskins (2004) defines mediating war through templates drawn from past mediations of crisis and disaster. Hence, the Summer 2007 UK floods drew upon media archives of flooding from 1947

---

[11] The Blitz is shorthand for that period (1940–41) during the Second World War in which the German Luftwaffe bombed major British cities.
[12] In his analysis of flood narratives Furedi calls for 'a more systematic engagement with the historical dimension of disaster consciousness' as an opportunity to 'illuminate the distinctive features of the contemporary response to adversity' (2007: 250). Furedi has continued to critique 'safety' and 'risk' in the context of catastrophes (such as earthquake, flood and more recently pandemic).

Figure 2.1 (above) St Peter and St Paul Church, Upton-on-Severn. (below) Boathouse on the River Severn (*Source:* Andrew Holmes). See our WordPress site for more images of flood marks and the associated interviews. Available at: https://floodmemories.wordpress.com/category/flood-marks/. Accessed 31 December 2018.

onwards, and as I have noted previously 'regional television news organizations would use their local and embedded knowledge to re-mediate their archival footage and repeat images over time' (Garde-Hansen et al. 2016: 62). It was through this re-activation of mediated memories that we delivered workshops at Gloucestershire Archives (now Gloucestershire Heritage Hub) during 2011–12 and drew upon our

Figure 2.2 Photo of the author taking photos of 1947 newspapers with a community group sharing memories of 2007 floods. *Source:* Andrew Holmes.

interview material[13] collected from communities to create new memory material for digital stories[14] as well as media sources from 1947 (print news and photography).

A typical example of such interview material remediated the Blitz spirit and corroborated the weather/war axis the journalist (above) had alluded to:

> A gentleman walked up to me and said in a German accent, 'Tell me, what is the difference between Germany and [Setting 1]? Four years ago we had massive floods and there was looting and fighting. I come to [Setting 1] and everybody has a smile, they're out sweeping the streets, emptying their houses of water. It's so different – why?' I said, 'It's the Dunkirk spirit!'
>
> (Male, 66, Setting 1, describing the floods of 2007 for a digital story)[15]

---

[13] In total, sixty-five residents were interviewed in depth across the case study areas over the three years, using snowballing techniques and a quota approach to sampling on the basis of gender and age to identify interviewees. The semi-structured interviews covered recording, communicating and maintaining or discarding flood memories, and their perceived relationships to community resilience.

[14] We curated a DVD of digital stories which was circulated by the Environment Agency to its flood risk and resilience groups throughout the UK, a copy of which is also held at Gloucestershire Heritage Hub under accession number D13866.

[15] In terms of anonymity, I have changed names in this book to basic details – for example, 'Male, aged 66'. I use a uniform/code description to reflect the regions of research.

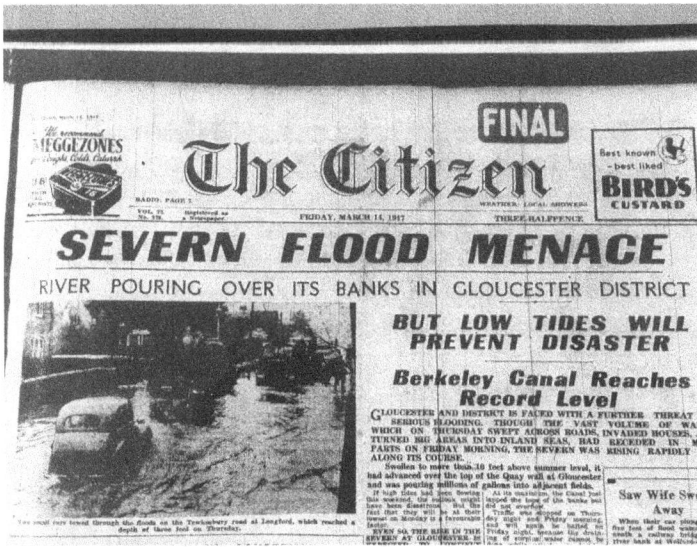

Figure 2.3 Photographs of two sources at Gloucestershire Heritage Hub. (above) Folder of 1947 and 2007 print newspaper materials which remains on open display at the hub. (below) Photograph of *The Citizen* newspaper headline from 14 March 1947. *Source:* Andrew Holmes.

The aim of the workshops was to enrich the interviews by bringing together the flood stakeholders (Environment Agency, local government, emergency services and resilience bodies) with the material produced from the workshops, created by what we termed 'flood memory agents' (Garde-Hansen et al. 2016: 61). Anna Reading has defined memory agents as those 'mainstream organisations or state and corporate memory agents' whose use of 'mobile and connective technologies [...] to "witness" such events' and connect 'prosumers' [producer-consumers] through 'trans-medial glocalised mobile connectivities and mobilisations' (2012: 23). In this case, the Gloucestershire Archives acted as a memory agent, and through their curation of 1947 print newspaper material (see Figure 2.3) and photographs[16] of the flooding we were able to draw together vertical and horizontal axes of remembering through media histories and alongside living memories as digital stories.

It is the curation of newspaper articles and images that became of relevance to the co-production of flood memory between participants and researchers. What is striking is that the participants were not only able to identify neighbours and friends in the 1947 photographic collection held by the archives (much to the delight of the archivist) but used the print media archives as stimulus for deeper conversations about flood histories and preparedness. This was an opportunity to connect flood memories from 1947 to flood resilience research after 2007. As we have noted previously, our participants 'were capturing, with the potential for sharing, flood memories through a range of modalities and with differing (even competing) but related motivations' (Garde-Hansen et al. 2016: 65). It was the use of digital media and the development of media literacy skills that these participants found liberating, as it allowed them to move the media archival material in new ways.

Case study 2: Media infrastructures of the 1953 Tempest

Storm surges[17] are mostly researched in the science and social science domains. We do have journalism mixed with science in Daniel Defoe's 1703 account of a cyclone landing on the Southern coast of Britain in *The Storm* (now a 2005 Penguin Classic), but there are far less cultural and media historians of the 1953 East Coast Storm surge[18] considering it is the flood event (following directly after the Lynmouth flood disaster of 1952) that is the most mediatized in BBC history until the 2000/1 and 2007 floods. It is also interesting to note that in searching for archival media holdings on

---

[16] Some research suggests that working with images can thicken interpretation, offering richer data, access to emotions, unconscious processes, beliefs or values (see Riessman 2007).

[17] Haigh and Bradshaw (2016) for NERC's Planet Earth series define a storm surge in 'A century of UK coastal flooding' as 'a temporary large-scale rise in sea level caused by strong winds pushing water towards the coast where it "piles up," and by low pressure at the centre of storms – this "pulls" the sea surface up by about 1cm for every millibar that air pressure drops'. 22 April 2016. Available at https://nerc.ukri.org/planetearth/stories/1812/. Accessed 12 December 2018.

[18] Matthew Kelly an expert in modern history has made the point in *The Conversation* (16 January 2017) that it was the East Coast Storm surge of 1953 that 'kicked off the debate on climate change' leading to the Waverley report, a new early warning system and the case for the Thames Barrier, passed eventually in 1972.

the 1953 Storm Surge one has to use the term 'Tempest' in the BBC Written Archive for evidence retrieval. This may seem outdated, but I would argue that the shift from encoding Shakespeare's play of the same name to a more recent meteorological discourse has emptied the event of some arts and humanities perspectives. This also means that historians of media are likely to pass over the evidence unless they wish to make the case that tempest, as a term, gives flood researchers access to old and new ways of understanding flood events.[19] In what follows, I will argue that both radio and television developed the capacity to take on spectacular[20] proportions but also to serve social, protective and charitable purposes, and the 1953 East Coast Storm surge saw all of these roles emerging. While the 1947 Winter Flood representation in print media and photography (as noted above) provided evidence for the construction and further refinement of a British collective memory of the 'Blitz' spirit (at least locally) even during a period of rationing, and extended the post-war preparedness for action, the 1953 Tempest provided evidence for media as a new kind of early warning infrastructure in a changing climate, reminding the audience of past floods as well as future concerns.[21]

It is important to acknowledge that British broadcasting history as defined by media and communication scholars has missed the importance of the East Coast Storm surge of late January and early February 1953 as pivotal to how the BBC developed in terms of policy, practice and technology. The general rule is that most media history books are concerned with social histories of politics, freedom of speech, democracy, power and representation (see Curran and Seaton's 2018, *Power without Responsibility: Press, Broadcasting and the Internet in Britain* now in its eighth edition). Scholarship on British media subdivides into discreet but connected histories based around the specificities of the media form, its infrastructure, production practices, modes of address and intended audiences: the history of newspapers in the UK at national and local levels or a history of the broadcasting of radio or television, for example. Thus, in *The History of the Provincial Press in England* by Rachel Matthews (2017) we have a page of analysis dedicated to a storm reported in the *Yorkshire Post* of 1895. While in Asa Briggs' official history of the BBC he mentions the 1953 East Coast Storm only once in the 900+ pages of Volume IV of *The History of Broadcasting in the United Kingdom: Sound and Vision* (1995).[22]

[19] See Georgina Endfield's 2017–18 Arts and Humanities Research Council project 'Weathering the Storm: TEMPEST and Engagement with the National Weather Memory' (with Sarah Davies).

[20] I draw radio into the 'spectacular' by considering Slavoj Zizek's point that: 'Voice does not simply persist at a different level with regard to what we see, it rather points to a gap in the field of the visible, toward the dimension of what eludes our gaze. In other words, their relationship is mediated by an impossibility: ultimately, we hear things because we cannot see everything' (Zizek 1996: 93).

[21] In the archival evidence I reviewed in the BBC Written Archives, radio broadcast scripts remind the audience of past inundations. Not only the floods on the East Coast in 1928, 1938 and 1949, but also the Lynmouth flood disaster of 15–16 August 1952 which would have felt very recent to the 1953 audience. It is worth noting that by the time of the 2007 Summer Floods the previous flood event had been 2000–1 and before that 1947 as the most extensive UK flood, quite a significant time gap for flood media to slip from memory. See Centre for Ecology and Hydrology's 'Major Hydrological Events' timeline for the UK at https://nrfa.ceh.ac.uk/occasional-reports. Accessed 12 November 2018.

[22] Briggs in describing *Home News* says it was following a 'familiar pattern broken only by *Radio Newsreel,* the success of which was obvious, particularly when it brought in the BBC's own overseas correspondents or when it covered brilliantly domestic events like the great East Coast floods of 1953' (Briggs 1995: 528). Not sure the floods were either brilliant or domestic thus dismissing their importance to the BBC's development.

Moreover, in all my years of teaching British media history to undergraduates I have not encountered core reading that covers environmental histories as significant to the story of the development of newspapers, film, radio or television. Rather, as Briggs shows, broadcasting was concerned with mastery, not only during the war in Britain but in all parts of the world, and 'as broadcasting had gained in influence and power, the influence and power of the BBC within the world network of radio communication had gained also. [...] known in Europe and across oceans [as] symbols in sound' (Briggs 1995: 26). When television's historical trajectory commences, it is 1952 with the Director General Sir Ian Jacob 'who presided over a great transformation' with a 'ten-year plan of development announced in 1953 – after the BBC's prestige reached its peak as a result of its Coronation broadcasts – looked forward ambitiously to large-scale future development' (Briggs 1995: 9).[23] Thus, it is far more likely that Queen Elizabeth II's coronation of June 1953 would become the starting point for British 'media eventhood'; one that brings temporality (the duration of monarchy), national identity (the phatic moment of collective memory) and media infrastructures (the live broadcast) into a single moment in broadcast history that shores up certain cultural values.

Yet an alternative history might be read through rivers, floods and water management. An overview of Genome (the BBC database of all *Radio Times* listings) shows a wealth of broadcast programming emerging in and after the 1952 Lynmouth Flood. Search results of broadcast material from 1923–33 (96 results) 1933–43 (98 results) 1943–53 (67 results) show a significant increase with 1953–63 (180 results), 1963–73 (433 results) and 1973–83 (386 results). The programming is international from the start with the 1920s covering floods in China, Egypt and America, the 1930s on floods in Poland, Africa, Brazil and India with increased capacity to report from abroad on floods in Poland, Germany, Holland, Hungary, Australia, Italy, Pakistan and Iraq thenceforth. These are alongside many programmes that focus on the biblical story of Noah and the Flood, mostly through children's programming.

The listings from 1952 onwards cover radio and television, with the 28 July 1953 memorial programme *Lynmouth – A Year after* broadcast on television and *In the Wake of the Floods* (25 July 1954) concerning the storm surge broadcast on radio. Both floods are regularly returned to through the broadcast history, often on or around anniversaries, such as in *The Ocean: The Great Tidal Surge of 1953* (24 June 1954); *The Invading Sea* (2 February 1954); *True Stories: East Coast Flood 1953* (3 February 1959); *Hard Weather: Flood and Inundations of 1953* (11 February 1963); *The Great Gale* (12 July 1961); *Plunder: Lynmouth Flood 1952* (13 August 1964); and *The Great Water Risk: The Problems of Flooding* (30 January 1969). From the 1970s the focus shifts from memorializing documentaries of past floods to fictional dramas such as *Troubled Water* (11 March 1969); *Midweek Theatre: Still Flows the Flood* (28 June 1972); *Play for Today: London Is Drowning* (27 October 1981); *When the Waters Came* (8 March 1982); or, *The Flood*, a three-part drama (12 February 1992). The timings of transmission (whether radio or television) to coincide with memories of flood events demonstrate

---

[23] Briggs notes that 'it was not until 1953 that the number of television sets being produced was greater than the number of sound receivers' (1995: 222).

an intimate intertwining of a watery sense of place and a media production schedule over a long period of time.[24]

Moreover, a growing environmental consciousness is detected in the written archive of radio scripts around this time, and this can get missed if we assume that only recently has science and social science drawn our attention to these issues in the light of climate change discourse:

> Each river has to be studied as a whole from source to sea. [...] It's up to man to do something about floods, because, you know, he's been a good deal responsible for causing them himself. [...] he's hacked down the trees standing on steep hillsides; the trees soaked up the rain water; now that the trees are down, they don't soak it up and it all pours down into the valley. And indeed you can see all over the country places where man has done something even stupider and worse than that; he has built new towns or town extensions in the flat low-lying bit of ground that was once the flood-area of a river.
>
> Floods, News Commentary, 8 February 1951

> Water affects the most important parts of our daily lives – our food, our clothes, our electricity – oh, almost everything you can think of.
>
> The National Water Shortage, Current Affairs II, 6 June 1956

While at the same time a sense of hope and possible denial about the freakish nature of such tempests can be found, suggesting that both remembering and forgetting can exist at the same time long before these intertwined polarities emerge as climate change believers and deniers:

> It's very unlikely of course that another huge storm of anything like the size of the one which flooded Lynmouth seven weeks ago will occur in the near future. The last one, as big as this present one, was in 1769, nearly 200 years ago [...] and it may well be another 200 years before 9 inches of rain falls again in one day on Longstone Barrow.
>
> Lynmouth, Current Affairs, 1 October 1952

> Of course, it was a freak storm, something that might not happen again in a hundred years or more.
>
> The Floods, News Commentary, 6 February 1953

> Quite suddenly the defenses against the sea in those areas were called upon to take the sort of hammering that they will not have to take again perhaps for two or three hundred years.
>
> The Recent Floods, Current Affairs, 11 February 1953[25]

---

[24] Only the more recent broadcast material is likely to still be available for broadcast as programme tapes were wiped and reused as a matter of BBC policy until the 1980s. Without listening to the originals it is not clear how much past material was fresh or re-broadcast material.

[25] With thanks to the research assistance of Dr Zöe Shacklock for helping to select these quotations from the BBC written archives.

Furthermore, research of the BBC Written Archives shows the BBC actively *working through* several important positions of communicative power. What is and should be the role of media in these extreme weather and flood events, in terms of radio and television production and broadcasting and its relationship to government and the Met Office and River Boards? What is the UK's relationship with other countries (such as in the Netherlands, Germany and India) with a shared sense of environmental consciousness of flooding and storm? On the part of the audience, there is evidence to suggest that the BBC was actively engaging, or at least thinking creatively about engaging the public in flood relief, adaptation and resilience. All of which foretells the rising importance of the Met Office and meteorology for both scientists of climate change and the general public in the twenty-first century. I will take three key ideas in turn and provide snapshots of archival evidence for them, suggesting that further research is needed if flood and water historians are to fully appreciate what media archives can offer.

a) Media as an early warning system

There is a great deal of paperwork in the BBC Written Archives during and directly after the 1953 Tempest pertaining to the UK government and the BBC in discussions about the role of the broadcaster as a potential early warning system for communicating flood events and flood risk. Apart from the wealth of science and tidal data among the broadcast schedules, there are these minutes from a 'Flood Warning Systems' meeting dated 9 February 1953 which discuss the BBC's position, and it is worth quoting the source at length:

> The Chairman said that the question arose whether the BBC should be requested during the next Spring tides to re-commence the regular transmission of hourly reports. [...] Mr Spalding said that during the past emergency these reports had been valuable to the River Boards because in these reports the information provided had been correlated and interpreted. [...] He agreed, however, that there were good reasons for the re-institution of the broadcasts; he suggested that the information given over the radio should refer to tide levels in relation to 'predicted tide heights' and not in relation to the previous tide height. [...] Commander Farquharson said that he advised that information for the public should be in terms of height of previous tides rather than 'predicted tide heights' because he considered that it was more comprehensible to them in that form. Mr Snagge said that the BBC would be prepared to act on the recommendations of the Committee. [...] He emphasised that the reports could only be of a general nature: the BBC could not undertake to broadcast instructions to the public in any localities. If the maximum benefit was to be gained from these reports he considered that they could take place at regular intervals. The only exception to this would be that the BBC would not interrupt church.

Such debates about the interconnectedness of media with water are borne out in the detailed outside broadcast (OB) instructions, memos and schedules that place BBC staff at key points on river banks and address the changes in OB engineering necessary in order to report on a flood event as it played out and in its aftermath. The confirmation of a BBC policy on early warning can be found in the file R34/1281 Policy Emergency Warnings: Flood 1968–71 in which a meeting takes place between Heads of BBC News (Radio), Ministry for Agriculture, Fisheries and Food about the following adjustments to the broadcast infrastructure: permanent line-fixed microphone to operational flood headquarters, early information to the BBC about London flood warnings, leaflets to Londoners on hearing the siren, to listen to Radio 1 (as well as TV and other radio channels); Radio 1 transmitters to be kept open after 2 am; and, finally, a warning system to be underway from 16 September 1968. The follow-up memo (21 August 1968) from the Engineer-in-Charge of OBs (F. E. Howard) details the technical changes to broadcasting and the drawing of technology from the Olympics in Mexico to be available for domestic broadcasting thus consolidating the idea of a flood as a media event.

b) Flooded nations connected through media

It is through media and culture that gifts are communicated between flood-stricken nations. A letter from Radio Nederland dated 30 March 1953 to the BBC states that the 'overwhelming and touching aid' provided to the Netherlands 'during and after the flood disaster' led them to create a bespoke radio production as a gift to the British people. Entitled *The Netherlands Thank the World*, they add 'We have the honour of presenting you with this version which we have made especially for Great Britain' (BBC Written Archives R47/297/1). The whole file 'Flood Programmes, RELAYS, 1953 R47/297/1' covers much of the background context to these new relationships between Great Britain, the Netherlands and Germany brought together around flood relief. Similarly, a few years later we have 'Experiences like this make you realize how close the people of Australia and Britain are to one another. Twelve thousand miles are really not so far at times like these; and Australians will not soon forget how quick the British were to respond when the great floods hit New South Wales in 1955.' (Floods in Australia, *News Commentary for Schools*, 16 March 1955). Thus, flood historians will find much evidence of transnational connections being made between media organizations as they develop practices for reporting flood and dealing with the aftermath and ongoing remembrance.

c) Media producers as relief, compassion and care

It is clear from the paperwork pertaining to the anniversary programme of the Lynmouth Flood in 1952 that the BBC learns a great deal about how to remember flooding in terms of ongoing care for the community, ensuring a wide variety of participants tell their stories and that the representation is sensitive to the relief and reconstruction efforts of local stakeholders (Letter to Devon County Council, 7 August 1953, Letter to Lynton Urban District Council, 20 July 1953, BBC Written Archives Television WE 9/18). That the BBC does not get it quite right is evidenced by the letters from the local councils and the Viewer Research Report which suggest the broadcaster needed to have more of the Lynmouth point of view even if the public were thankful for the clear explanation of the flood. Following the East Coast Storm and flood of 1953 the files suggest a change of role for the BBC. There is the '1953 Lord Mayor's National Flood and Tempest Fund' established (found in a Talks/Appeals file T32/387) and with this brings discussions within the BBC about how to manage all the donations coming in from the public such that they establish an appeals section. There is the International Flood Relief Show at the Royal Albert Hall broadcast on 1 March 1953 coordinated with a broadcast in Holland the same evening and a discussion of a giant cake to be auctioned on a television programme idea (though it appears this programme was stopped by senior BBC) as a new way of engaging the public in flood relief through community spirit and entertainment. By the late 1950s the files suggest a shift in flood media towards education bearing in mind Hugh Carleton Greene's speech to BBC staff on 12 March 1958 in the wake of commercial television: 'after all the children of today are the adult viewers of tomorrow' (BBC Director of Administration, BBC Written Archives). By 1969 we have files such as 'Education Schools Programme: Exploration Earth' (R16/913/1) wherein the Schools Broadcasting Department details how it intends to incorporate flood education into a programme with a factual geographical basis and with direct references to real centres of damage. All of which suggests a closer appreciation of broadcast archives can help researchers understand a much longer history of responsibility and resilience.

Conclusion

Broadcast archives and their written material are rarely used by media researchers of climate, weather and environment who do not see in media history an integrated history of environmental and natural resources; and water researchers who often prefer to focus on more recent media reports and only then as a small part of a larger project. For example, in their 2015 report *Public perceptions of climate change in Britain following the winter 2013/2014 flooding*, a research group at Cardiff University, UK, noted that flood events receive 'significant national and international media exposure' (Capstick et al. 2015: 7). The respondents to their questionnaire on the relationship between flooding and climate change (which included examples from news media reports on recent flooding) indicated they became more concerned about climate change itself when they experienced 'directly observable weather phenomena' [69 per cent] rather than 'media reports' [13 per cent] alone (Capstick et al. 2015: 23). They concluded that there is

> a significant association between the winter flooding and climate change [that] has already been formed in the British public's mind. Perhaps this should come as no surprise, given the fact that climate change and UK climate impacts have been rising on the UK environmental policy agenda for well over 15 years now, while at the very same time incidences of major flooding have become a recurrent topic of British media attention stretching at least back to the major flooding in York in November 2000.

(Capstick et al. 2015: 44)

This is certainly true and has been the case for much of the history of the print newspaper industry in the UK (from the nineteenth century onwards) and certainly for the last one hundred years of broadcast media history, with the 1947 Winter Flood, the 1952 Lynmouth Flood and the 1953 East Coast Storm as striking examples that grapple with two key issues: how best to represent flooding and prepare the public for future events; and, how media production, practice and infrastructure can serve that purpose. The case for daylighting broadcast archives is based on the need for more scholarship concerning contemporary flood events and the wide sharing of experiences of flooding through mainstream and social media that have been seen by science communicators as the entry points for a larger climate change communication strategy. However, it is only through a closer and deeper analysis of specific types of media in their historical contexts of production that we can make any substantive claims about communicating water risks to the public, and in a way that (re)connects the public to their own mediated water histories and inherited memories.

3

Socially mediating water for digital hydro-citizenship

Storms of social media

The previous chapter made the case for analogue media archives in researching flood and storm events, and the rich resources of largely unsearchable (in terms of digital search), typed and handwritten texts that surround the print and broadcast media of the twentieth century. They are important for developing a context specific, historically located and evidence-based approach, which develops the connections between media, culture and communication to 'hydro-citizenship'.[1] The latter is defined by Owain Jones in his research as:

> How local communities are embedded in the hydrosphere (the totality of interconnected water forms, cycles, systems, issues, conflicts), and using arts and humanities centred interdisciplinary research (AHIR) to explore and develop community resilience in eco-social terms. [...] addresses multiple, interrelated water issues (floods, drought, water quality, biodiversity, ground water, catchment management) considered within community contexts. Interconnected water issues are some of the most challenging in the UK and beyond, particularly, in the context of climate change.

(Jones 2018: [online])

Thus, media archives of the twentieth century may appear to non-historians in a digital age as bureaucratic, hierarchical and organized around provenance (i.e. the programme, film, director, writer, key location or event) rather than community or theme focused. Yet while these archives are not easily accessed offline or online and rarely open to the general public, they are dormant and ready to engage audiences in community-based conversations.[2]

---

[1] See Jones and Jones (2016) 'On Narrative, Affect and Threatened Ecologies of Tidal Landscapes' and Payne, Jones, and Jones (2015) 'The Hydrocitizenship Project Celebrates World Water Day 2015'. Also the video 'oath' to the Rivers at https://www.youtube.com/watch?v=u79gnQD39x8&t=178s.

[2] There are a handful of semi-accessible UK media archives beyond the BBC. The Media Archive for Central England (MACE), and the National Library of Scotland, the commercial archive of ITV and Ulster TV archive, as well as the BBC Motion Picture Gallery/Getty Images housing images are examples.

By daylighting evidence from media archives and seeking more access to them we can see a 'cultural memory' of water recirculating as potential 'communicative memory' in a digital age. This suggests that one of the most relevant typologies for socially mediating water through new forms of communication networks is from the memory research of Jan Assmann (1995: 128–9) in his distinction between 'communicative memory' and 'cultural memory' to draw attention to the new relationships between personal or community mediations of water and official, institutional and organizational repositories and archives. In this chapter, we move, then from the cultural memory of broadcast archives to the communicative memory of social media platforms and the uses of these by water companies and organizations. Therefore, inspired by Myerhoff's (1982) work, this chapter involves recollection of a different sort, by analysing social media as a communication tool that seeks to 'thicken' water stories by connecting them (through platform power[3]) to people, places and practices at the personal and local scale, while having global reach. Such an approach forms a central challenge for science and social science researchers of water issues in the twenty-first century. As Clandinin noted in the *Handbook of Narrative Enquiry: Mapping a Methodology* (2007: 1) there have been four key turns in the movement of narrative enquiry: 'a change in the relationship between the researcher and the researched; a move from the use of number toward the use of words as data; a change from a focus on the general and universal toward the local and specific; and a widening in acceptance of alternative epistemologies or ways of knowing'.

Consequently, in a rapidly changing media landscape of digital technologies, the issue of who gets to tell their story of water and what the story covers remains critical even if those technologies are changing the language and the structure of social connections and scales. Beyond one of the five '-scapes' (Appadurai 1996)[4] we now have a much more diverse 'mediascape', one that Geiß (2018) suggests is structured by news media through 'eventscapes' of 'genuine events', 'mediatized events' and 'staged events', which can be defined in terms of whether they are 'random events', 'serial events' or 'recurrent events'. A flood, storm or drought can fall into any of these latter three categories and that will be determined by the ability to connect water stories through a longer narrative of climate, access to expertise from the science community, access to local knowledge and the seasonality of such events in a particular location. But clearly, the connectivity of media across scales, genres, practices and discourses means there are now multiple producers and audiences (of one or many) who have renewed capacity to take part in meaning-making processes, co-producing stories of

---

[3]   For more on platform power see *The Platform Society: Public Values in a Connective World* (2018) by José van Dijck et al. They use the word 'environment' in the sense of platforms creating environments as part of a wider platform ecosystem. The key issue is that platforms (most based in the United States) are paradoxes: five high-tech companies dominate (Alphabet-Google, Facebook, Apple, Amazon and Microsoft) though I would add the Chinese companies Tencent, Weibo and Huwaei are rising powers. They look egalitarian but are hierarchical, they offer public value but are corporate, neutral but ideological, they appear local but are global, empower the consumer/user but are centralized and opaque say van Dijck et al. (2018: 12).

[4]   Appadurai (1996) covers '-scapes' of global cultural flow: ethnoscapes, financscapes, technoscapes, mediascapes and ideoscapes; and he covers the 'scalar dynamics' of globalization in Appadurai (1990); and he covers the 'scalar dynamics' of globalization in Appadurai (1990).

water as online conversations, while also sharing and listening to stories outside their own experience and contexts. One should be wary of moralizing discourse that seeks to shut down social media's capacity to connect audiences in outrage, especially if that outrage is directed at shameful environmental practices by powerful players with teams of marketing and public relations officers. Similarly, one should be wary of the use of social media to hijack personal and local water concerns for ideological point-scoring. If one only sees a tweet, photo, comment, status update or even a selfie in a waterscape as the story, then one misses an understanding of social media itself as a *genre* that is accumulating micro-stories into serial narratives over longer time spans. We must pay attention to the narrative arc and not only the momentary story.

Consumers become producers argues Henry Jenkins throughout *Convergence Culture: Where Old and New Media Collide* (2008)[5] and with their demand for participation in storytelling we have an 'evolution of social interconnections', 'a consequential blurring of boundaries between public and virtual space' and a need to 'encourage processes of mutual understanding among widespread communities of interest and practice' (Ciancia et al. 2014). Moreover, as noted in the previous chapter, media infrastructures matter, and Clark et al. (2014: 924) argue that:

> Not only may digital narratives take many forms, including shorter forms (tweet, retweet, blog, posting an image or video); the question of how the interaction of digital forms and processes creates over time a wider *infrastructure* through which narratives are sustainably exchanged becomes increasingly important.

Therefore, it makes sense to see social media as more than an opportunity for everyday knowledges on environmental issues to be shared widely, as do Endfield and Veale (2018: 6) on weather events, in which contemporary 'instances of unusual or extreme weather are now perhaps more likely to be captured and recorded through webpage entries, blog narratives and tweets as they are newspaper reports or other written records'. Rather, social media challenges us to tell stories of water beyond the text,[6] in very small ways that cannot be simply 'weatherized' as storms and waves of news and opinion. Social media platforms offer opportunities to daylight the quotidian, to see the ordinary feelings about water and to create narratives and counter-narratives of water management, which may bypass the filtering power of professional storytellers.[7] As Krause et al. argue (2014: 8, 10, my emphasis) in their research on the use of Twitter by non-media organizations:

---

[5] Jenkins argues in *Fans, Bloggers, and Gamers: Exploring Participatory Culture* (2006: 60) that 'consumption becomes production; reading becomes writing; spectator culture becomes participatory culture'. One could argue, for example, that certain rivers have fans and followers.

[6] This can be connected to the inventive methods that Beebeejaun et al. (2013) reference in their research. Narrative methods that they see as 'beyond text' – including photovoice, photo-elicitation, social media, 'new text' methods, film-making, storytelling, blogs and presentations. This is part of wider shift in de-colonizing methodologies of research to be more inclusive of participatory and co-produced approaches that are not so logo-centric.

[7] This is a contentious issue and I am not advocating the redundancy of journalists. Quite the opposite, like science communicators they are going to have to work much more collaboratively with audiences on topics of science, verifiable knowledge production, data and story-making.

Twitter has *stormed* onto the social media scene not only as an individual communication device but also as an information dissemination platform in times of disaster. [...] What reporters and government entities have in common is that, over time, the information of individual disasters, crimes and other public events coalesce on Twitter as mini-stories for public consumption.

All of this suggests that social media genres work best when place-specific, empathic, creative and integrated into everyday life but that does not mean as researchers we should remain uncritical of the local and lay knowledge-based stories of water being circulated. For climate change communication it is key because as Lejano et al. (2013b) argue, a 'narratological approach' (starting where audiences are as the receivers, readers and creators of their own life narratives) allows climate discourse to adapt to and integrate into everyday lives 'to become more commonplace and not isolated from other issues' (2013b: 61). They continue:

> Unless experts and policymakers are able to craft or engender stories about things like climate change that integrate this issue coherently with other aspects of people's everyday lives, such as our identities, beliefs, and experiences, these issues do not become salient enough to the public.
>
> (Lejano et al. 2013b: 62)

However, there is no point arguing in favour of community knowledge of and community education around water (Allon and Sofoulis 2006) if we do not also acknowledge that the interrelation of water (or lack of it) and media brings a community into being in the first place. As noted in the previous chapter, water and media have been connected from the get-go and communities can be built upon their water stories in whatever form they are delivered. Thus, one cannot assume that a stable and identifiable community, family, neighbourhood or nation prefigures media, particularly if that community's social media scape is being defined in terms of troubled waters.

Like any storm, then, a Twitterstorm[8] spikes in key places, then plays out in a scattered manner, connecting up disconnected actors, spaces and places, perhaps triggering or being triggered by a 'news wave' (see Vastermann 2018). Like all waves it has a shape, as noted by Geiß (2010) in 'The Shape of News Waves' where there are 'slow burners' that escalate, 'firestorms' of news, short 'news waves' that 'can be explosive with a long tail, or, conversely, start slow, but explode later' or the '"degressive" or "progressive heating news wave"' (Vasterman 2018: 21). All of which suggests that the social media environment is currently being conceptualized through troubled water and weatherized as an uninhabitable environment, perhaps with only

---

[8] 'A Twitterstorm is a sudden spike in activity surrounding a certain topic on the Twitter social media site. A Twitterstorm is often started by a single person who sends his or her followers a message often related to breaking news or a controversial debate' (www.techopedia.com) but it is also often unintentional suggesting that a Twitterstorm can erupt anywhere. Moreover, the capacity of the internet to remember the storm is high if new kinds of stories and narratives do not work to deliberately push the activity spike below the surface.

traditional 'news' as a saviour and elite persons as the guardians.[9] Such an extreme weather metaphorization of social media in the twenty-first century creates a sense of risk and fear, expects resilient storytellers and plays into the hands of authorized, professional and corporate media who may remain opaque. What happens in between the storms, spikes, waves and the ebbs and flows of news media and social media? Is it only through these kinds of media dynamics that water becomes memorable, sets the agenda, surges attention and creates a 'media tsunami' (Giasson et al. 2010)? Outside of the news agenda, how can social media genres offer opportunities for old and new media on water events and water's representation to collide? Where can we find inhabitable and collaborative spaces for socially mediating water and by extension draw water and media into a conversation for a less troubled environment?

In what follows, I address the increasing use of social media by water managers, by water users to engage water companies and in water campaigns on issues of water use. This sets the scene for Chapter 6 on digital memory and flood-risk management and Chapter 8 on riparian media and culture. Tweets about leaks,[10] using GPS and social media to direct water engineers to problems; texts to phones updating on water shut downs which include links to online stories of water management issues; the use of video and photographic media to engage the public on water engineering or water scarcity; the sharing of water stories and the gamification of water usage as well as hashtags; social media campaigns by governments and NGOs; and creative fan fiction that asks us to rethink water in the city. Together these provide new and emerging everyday circulations of mediating water. I will argue here, and in later chapters, that social media has the capacity to widen the administrative levels, connect the hydrological scales, increase the actors and the networked relationships, their roles and social media capacities, and reveal the water conflicts as well as compromises. In the hands of professional storytellers (customer engagement teams, media influencers, public relations and marketing) it is a powerful communication tool and there is plenty of evidence of the emergence of what I will call *digital hydro-citizens*[11] who are taking up old and new roles as water communicators. The analysis begins with original interview material from the water industry on managing media in a flood event and culminates in thinking through drought @mentions and drought shaming on Twitter.

---

[9]   In *From Media Hype to Twitterstorm* (2018) Vasterman's collection uses the term 'wave' on 179 of the 403 pages, several times on most of those pages, alongside the words 'stream', 'tsunami', 'flow' and 'deluge'. The whole collection is 'weatherized' such that the 'news seems to develop a life of its own *like a resonating bridge in the wind*. And it is not only the news, it also applies to the public arena as a whole, including – in this digital era – internet and social media' (Vasterman 2018: 18, my emphases).

[10] For example, the San Diego County Water Authority launched a mobile app for users to report water waste issues in times of drought. Available online at:http://www.watersmartsd.org/news/water-authority-launches-mobile-app-help-report-and-fix-water-waste-countywide.   Accessed 1 September 2016.

[11] Taking Jones's definition of hydro-citizenship, I have adapted the concept with researchers in Brazil to encompass participatory cultures and social media wherein we are exploring 'the connectivities between stories, narratives, anecdotes and digital mediations of water in São Paulo state and the UK. The focus is upon digital communication, social media conversations and interactive media cultures' (see the Narratives of Water project 2015–17 at https://narrativesofwater.wordpress.com/).

Reporting water: 'There's no script for this'

With large-scale issues, such as water events, water use and water misuse, reaching informed conclusions on how to approach reporting water is challenging. So says journalist Lis Stedman (2014: 125) because of the water sector's 'sheer size and complexity', requiring interdisciplinary thinking and 'a delicate balancing act' between expertise and lay knowledge. I would add, reporting water is also challenged by the tensions between privatization and the retreat of the state as well as the pervasiveness of neoliberal logics that places water in private hands (see the UK's Water Act 1989), and responsibility for its communication at atomized levels. Privatization may have been resisted in Australia (see Edwards 2013), for example, but tensions exist in many countries between globally marketed environmentalism, state interventions, collectivist concepts of nation-building and the growing power of 'Big Water' as defined by Allon and Sofoulis (2006). Thus, in the context of neo-liberalism, Foucault would argue that being entrepreneurial (as a journalist, a water manager, an environmental campaigner, a policymaker, concerned citizen or as the solitary social media operator of the water company's Twitter account)

> involves extending the economic model of supply and demand and of investment-costs-profit so as to make it a model of social relations and of existence itself, a form of relationship of the individual to himself, time, those around him, the group, and the family.
>
> (Foucault 2004/2008: 242)

Wynne suggests in 'Strange Weather, Again: Climate Science as Political Art' (2010: 299) that 'mainstream social science tends to reinforce an atomized and instrumental, rational choice self-interest model of the human subject'. Thus, science communication on water issues and water-related events often draws upon what I would term the marketing and communications end of the spectrum to engage the public rather than the creative arts and humanities[12] and in doing so, it makes an equation between open science and open media as providing mutually beneficial windows on understanding water in the environment and economy.

Not all media infrastructures are open, democratic and with a free press. Underlying the communication of science is a belief that media (and increasingly social media) play a central role, and that with the right kind of messaging the public can better understand the 'current water challenges, the reality of the water cycle and its technical requirements, its complex legal system or pricing' and communication could be based on 'the scale, the target groups, wording and visual identity' and the 'long-term vision and strategy employing different kinds of campaigns' (Hervé-Bazin

---

[12] Climate Communication.org uses animated and narrated graphics to express temperature rise. Vimeo is used extensively to upload high-quality engaging stories of water scarcity. For example, *Dry and Drier in West Texas* concerning droughts from 2010 onwards by the Colorado-based The Story Group (described as multimedia journalism) at https://vimeo.com/thestorygroup. In 2018 the *Just Water Challenge* for Water Aid received extensive coverage online and through social media networks.

2014: xv).[13] Which begs the question as to whether the public wish to have *their water* (its personal, local, national and global story-ing) *marketed to them* through long-term visions, scenarios and strategies, and whether corporate communication strategies are best placed for *engaging audiences*? As science communicators, policymakers, NGOs and governments take up social media channels for communicating water issues[14] this should not assume that media and communication frameworks are neutral and that audiences are not already highly media literate and immersed in their own mediated waterworlds.

In 'Muddying the waters or clearing the stream? Open Science as a communication medium' Grand et al. (2010) draw media communication and watery metaphors together to address the open-ness of science and data with an increased capacity for flow and transparency in a digital age. To use blogs, engage citizens online, or create open source data, can engage 'a variety of participants beyond the research groups that generate their content, ranging from members of the public to other professional audiences'. However, although 'direct access could "clear the stream" of communication, the sheer quantity and complexity of the process and data could "muddy the water", rendering navigating, interpreting and analysing the available information difficult' (Grand et al. 2010 [online]). Reporting on water, then, has to be understood in the media and communications research field as framed just as much by ideas of 'good' or 'bad' media (clear or muddy) as by 'good' or 'bad' water.[15] Thus, to draw on the ideas of Krause and Strang in 'Thinking Relationships through Water' (2016: 634), to socially mediate water would be to theorize a relationship between media and water wherein both are idealized and projected onto (as surfaces) by professional communicators as well as public audiences. Both are conceptualized as screens, landscapes and stages on, in and through which human actions and politics play out. Yet, as Wynne (2010: 299) notes, what of the 'habituated practices', the micro-stories, the potentially erased memories that may become backgrounded within a media ecology in which socially mediated water stories circulate every day. I am not thinking here of the celebrified social media accounts attached to key figures in water communication who appeal to ordinary citizens, such as Canada's *The Water Brothers* (https://twitter.com/thewaterbros) who have undeniably had an international impact with their water adventure series and associated campaigns, talks and environmental actions.[16] Rather, I am focusing on the

---

[13] In *Water Communication* (2014: xxiii) Hervé-Bazin says there are five main research areas for water communication: social representations of water, water discourses, knowledge brokerage, water campaigns and water journalism.

[14] The following have had media campaigns, logos, brands and public relations professionals setting agendas: WWDR4 (2012) *Managing Water under Uncertainty and Risk: The United Nations World Water Development Programme*; Water Wise (2013) *International Year of Water Co-operation*; UN Water (2013) *Water factsheets*; UN Org (2013) International Decade for Action 'Water for Life' 2005–15; Mayor of London Assemblies on 'water' 2011–17; and the *EU Water Framework Directive* (2013).

[15] See the campaigns 'Water Kills' (2007), 'Water Wall' (2010), 'Water Ink' (2011) and the 'Minute' (2012) of the French NGO Solidarités International who argue in favour of complete simplicity in messaging (case study in Hervé-Bazin 2014: 127–8).

[16] See their website at: http://thewaterbrothers.ca/.

hyperlocal perspectives and snapshot conversations between water companies and ordinary consumers/users, which are not part of a media eventscape.

This is where leaks and leakiness take on new dimensions.[17] In recognizing the ideological construction of water as 'blue' versus 'brown', as noted by Linton and Budds (2014), we find a similar muddied or transparent media and communication during the 'furore surrounding the *leaking* of emails from the Climate Research Unit at the University of East Anglia, UK' (Grand et al. 2010, my emphasis). In an era of 'fake news', climate sceptics and alt-right journalism, leakiness has been taken up with valour and so science communication has entered a media ecology where science has found itself accountable. An open media infrastructure is just as leaky for a large private water company as it is for the powerful voices of well-funded research communities and national media organizations. Thus, in the face of a water crisis, how should a private water company provide good water through good media using good communication skills? Through the simplified messaging of a company's marketing team and public relations or through a variety of social media genres?

In an interview with a former operations manager of Severn Trent Water in 2013, reflecting upon his own and the company's reporting and media interviews about the Summer 2007 Floods, he makes a sharp distinction between corporate communication and personal communication. In effect, this shifts the responsibility for communicating a water event to the individual member of staff overshadowing the fact that after the flood Severn Trent was discovered to have seriously misreported its own communication of water issues to the regulator, OFWAT.[18] Placing himself in the present tense of 2007, he states it is not a time for 'corporate nonsense' (a euphemism for marketing speak or faceless public relations); we 'cannot afford to put a PR person up on the cameras' and so 'I concentrate on the main media outlets' which 'gave a continuity', a narrative, but the 'minute a story becomes big, you are *swamped* with demands for interviews, comments, and it takes some managing' (interview 10 June 2013, my emphasis).[19] In fact, he says, to report on the floods was isolating: 'it can feel very lonely, you are sat there, you are asked to speak on behalf of a company; the expectation of the company, quite rightly, is on you, and they're watching it; you've got all those staff, which I thought more of, out there, relying on you to come across in a way that gives the viewer confidence, as that's the customer, and *there's no script for this*' (interview 10 June 2013, my emphases). In 'The Neoliberal Self', McGuigan describes

---

[17] See definitions at www.wikileaks, www.medialeaks and the many social media campaigns that seek to leak information in the face of non-disclosure agreements.

[18] Severn Trent PLC is a private water company (see https://en.wikipedia.org/wiki/Severn_Trent) and derives its name from two major rivers in England, the Severn and the Trent covering Birmingham, the Midlands and the South West counties of England. After the 2007 Summer Floods the company received hefty fines from the regulator OFWAT. Details of all the fines imposed on Severn Trent for misreporting on leakage can be found in the UK National Archives database.

[19] He undertook interviews with all the national and international news channels during the period of the flood event. It should be noted that local media were the main focus of audiences. In research on public health communication of water quality during the floods Rundblad et al. (2010: 11) concluded that 'Radio Gloucestershire was able to establish themselves as a timely and trustworthy information source. Building on the example of BBC Radio Gloucestershire, it is essential that local media be pre-prepared and, during an event, be continuously updated to maximise their role in ensuring public safety.'

'the individual [who, if things go wrong] is penalised harshly not only for personal failure but also for sheer bad luck in a highly competitive and relentlessly harsh social environment' they are 'condemned to freedom and lonely responsibility' and this 'is exactly the kind of self, cultivated by neoliberalism, combining freewheeling consumer sovereignty with enterprising business acumen' (2014: 234).

Hence, the entire interview hinges on the water manager's belief in transparent communication concerning a flood and water scarcity emergency, with care and compassion for those with sewage running through their homes, for a major incident of hundreds of thousands of homes evacuated and regions without clean water, carefully positioning himself as spokesperson for but in contradistinction to the values of the company. The flood caught the regulator and the industries 'on the hop' he claims which was the 'biggest lesson' but the second lesson concerned mediating water:

> More specifically on media and customer relations [...]. It reinforced the belief which many people have, that if you tell the truth, however bad it is, that people have a *pre-resilient appetite and ability* to deal with it, and actually, if handled in the right way, in general terms, you can get them on side and you get them to help and work with you. The minute you try to go down some route of corporate talk and prevarication, you lose those people, and it's virtually impossible to get them back. That didn't happen because we told the truth from the start [...]. In actual fact I got a letter of commendation from the Director General of the BBC, which I've kept in my file at home because I was quite proud of it, which commended me, and also the company, for the way in which we engaged with our staff [...]. We were so open and honest with viewers and listeners.
>
> (Interview 10 June 2013)

As the mediator of a water event, the Severn Trent Water operations manager acted not only as the mediator of emotions but also as the channel through which stories of natural disaster flow. Once Twitter arrived later that year such channels became steam valves as the democratization of media removed the gatekeepers.

Letting off steam online

The water manager's framing of Severn Trent's approach to the communication of the 2007 Floods, just prior to the roll out of Twitter across many organizations in the UK, may appear to have continued in their branding, communication strategy and the design of their fleet of vehicles.[20] The company seeks to project an image of care, culture, history and place, and managing media relations (like a dam manages

---

[20] The company represents itself visually in the streets through its vehicle livery as engaged in culture, heritage and social issues. There are various vehicle designs but all have a large photographic image on the side and a slogan; such as 1) image of work inside a massive pipe, 'Making it Better' 2) image of engineered pipe, 'Hidden Treasures' 3) image of hand trickling water onto a seedling, 'Wonderful on Tap' 4) Image of workforce at night, 'Extra mile'. All are designed to foreground cultural and social values rather than a more complex story of water privatization, regulation and misreporting offences.

the reservoir) is key to stemming the flow of bad publicity. However, that does not mean communicating with the public on water issues is easier in a socially mediated everyday encounter. In fact, I would argue that platforms have taken on new watery metaphors for public anger, shame and discontent. Thus, the public can use platforms as a 'steam valve' for 'citizen anger' (De Lisle et al. 2016: 2) or 'safety valve' (MacKinnon 2008: 33) to release resentment at incompetence but also the water companies can use the platform to identify potential problem water stories and stem the flow or channel these in order to maintain social stability around their water narratives.[21] All of which happens outside and beyond mainstream and traditional local and national media, for example, the public sees water leaks and it leaks bad water stories on Twitter.

Consequently, it is now commonplace for Severn Trent Water to use Twitter[22] to communicate with its customers on everything from staying hydrated during New Year's Eve parties to the building of a reservoir, from how to dispose of oil and fat to photo-by-photo updates on the repair of a burst pipe. They are acting in the public interest, nudging behaviour, communicating social values and collecting information. The vast majority of the Twitter account's *Tweets&Replies* are one-to-one exchanges with an individual customer on a domestic water issue: brown water from the tap, frozen pipes, water leak, water shut off, etc. Most of the photos uploaded are of water workers in all weather conditions and at all times of day working industriously to replace brown, dirty, ageing and clay pipes with new, blue, plastic ones, while the reality in 2020 is that 3 billion litres is lost through leaky pipes every day. In one such tweeted video, which produced a social media conversation of eight comments, eighteen retweets and thirty-nine likes (considered a quite active tweet in this context) a reported leak in a road in Birmingham produces *Replies* which turn Severn Trent's intended water narrative into a counter-narrative. Unfortunately, Severn Trent would not grant permission to include a screenshot of the tweet for analysis, and when probed on this they said they never grant permission to authors to reproduce a tweet and its replies in any print publication for any purpose whatsoever. Therefore, the reader can track down the tweet (assuming it has not been removed) from my descriptions in what follows.

The tweet posted by Severn Trent Water's handle @stwater on 13 December 2018 consisted of a 25-second video of a burst water main in Birmingham with fluorescent-workwear-clad Severn Trent engineers fixing the problem in a spectacular waterfall of ascending water and spray. The video is framed by Severn Trent presenting the engineers as local heroes, clearing the problem in a busy rush hour spot, garnering 3,381 views and a handful of comments which soon turns into a water discussion. While some comments applaud the bravery and hard work or feel sympathy for them

---

[21] The 'steam' metaphors are used primarily in research on China's social media platforms. See, for example, Hassid's (2012) 'Safety valve or pressure cooker? Blogs in Chinese political life'. One should not assume that research on social media used in an authoritarian Communist party state is not applicable to democratic nations. In fact, media researchers may be looking in the wrong places for understanding how social media flows are being powerfully controlled in a digital age.

[22] Many UK water companies, at least in the privatized English sector, use social media influencing. See Thames Water topping the charts according to 'Water and Wastewater Treatment' (2017) article https://wwtonline.co.uk/news/thames-water-tops-water-company-social-media-list and tracking platforms such as RISE are keen to expand their 'utilities gamification' model for water companies in the UK https://www.rise.global/water.

getting so wet (thus promoting water work in the public consciousness), two unique *Replies* successfully derail the narrative intended by the company:

> *Reply 1:* You charge us a bloody fortune in water rates, which we have no choice but to pay and you're trying to fix this with 2 blokes and a shovel whilst your shareholders soak up the profits. Bloody marvellous. Privatising monopolies is great isn't it. 😡
>
> *Reply 2:* When are you going to fix the leak on Wake Green Road near the junction with Springfield Rd? It's getting worse every day and will cause a downhill skid pan in this cold spell?
>
> *@stwater: Hi Barry, I'm sorry to hear about the leak, I can look into this for you. Do you have a postcode and property number this is showing outside of? Thanks, Andy*
>
> *Reply 2:* Sorry, no. I have driven over it twice a day for more than a fortnight. It is breaking up the road surface and filling potholes with water.

It will depend upon one's politics concerning privatization of England's water companies or one's level of sympathy for the Twitter influencer caught off guard by an alternative water leak story (when trying to narrate the company's image), in determining what the water reality is at this location.[23] As Strang notes:

> Interactions with water take place within a cultural landscape which is the product of specific social, spatial, economic and political arrangements, cosmological and religious beliefs, knowledges and material culture, as well as ecological constraints and opportunities.
>
> (2004: 5)

If cultural strategies including films, photos, digital stories, text and emojis are being converged and used to communicate 'key messages' relating to water management at micro-scales by macro actors (such as water companies), then we need to also pay attention to the micro counter-narratives and ordinary stories these strategies smooth over but never quite succeed in stemming the flow of.

The upturning of the one-way flow of information between professionals and publics by using culturally relevant and provocative methods and encouraging dialogue between water managers and communities as Berryman et al. claim (2014) does not fully erase the habituated practices of hyperlocal stories and experiences. What is key here, in this extremely small sample which is representative of the fairly regular water consumer conversations with Severn Trent PLC on Twitter, is that consumers are engaged in a daily and iterative form of digital hydro-citizenship about their rights and responsibilities. They are seeking answers, assurances, information and updates in real time; they care about their urban environment; they are quick to shame the water company; and they anchor this in photos and videos that report water in a myriad of ways (leaks, reservoirs,

---

[23] See Bruns and Burgess (2011) 'The use of Twitter hashtags in the formation of ad hoc publics' for more on the sustainability of these publics and Murthy and Longwell (2013) 'Twitter and disasters: the uses of Twitter during the 2010 Pakistan floods' and Jansen et al. (2009) 'Twitter power: Tweets as electronic word of mouth'.

hydration, pollution, engineering works, scarcity). The public are also challenging the dominant narratives of the professional storytellers (water company public relations departments) and the steam valve of social media clearly shows evidence of protest, eco-awareness and atomized individuals holding power to account in everyday ways.

Emotion online: Drought-shaming

The public's pre-resilient behaviour, their need for transparent communication and the emergence of a digital hydro-citizenry all suggest that professional water communicators (science, industry, policy) may be misunderstanding the public without a fuller and deeper appreciation of the social mediatization of everyday life that has concerned media scholars for the last decade or so (see Livingstone 2009). Science communication does not always see 'the public' as a media-literate audience already engaged in water issues but assumes a deficit model. That is, the public as conceptually consumers within a private-public dialectic of the market for a natural resource; or as water stakeholders who need perceptual, attitudinal or behavioural-correction in order to take up their proper roles as custodians of their local (and increasingly globalized) environment. In this sense, 'the public' can be said to be 'responsibil-ized' by science and climate organizations communicating media messages of water scarcity, risk, pollution and extreme weather adaptation while at the same time that public demands access to a seemingly abundant resource without a proper understanding of drought, for example. The public, through communication of participatory water frameworks, may become socially tethered to watersheds, river basins, local watercourses and resources as stakeholders yet find themselves at the mercy of more powerful stakeholders or made responsible for a resource they feel strongly has a universal application.

In any case, the public or even the citizen may be conceptualized as an empty concept, one that needs governing, with little real knowledge concerning water issues; to be properly educated, informed or made responsible. Thus, it is not always recognized that members of the public are already deeply engaged in water issues, inhabit water storyworlds of their own and do cooperate or wish to collaborate with water companies and environmental agencies even if they find their own voices, stories, memories and mediations of water largely hidden from view, or channelled through a consumerist or conflict discourse. They may also believe strongly, based on local knowledge, family stories, personal experience or via social movements and networking that they do not wish to become part of a new kind of 'water regime', which may be participatory but leaves individuals feeling solely responsible for water. Defined by Bakker (2001: 156) in the UK context of metering water during the Labour government, a 'water regime' signals a neo-liberal shift from rights and entitlements to water, to the 'basic needs' approach.[24]

---

[24] Bakker notes: 'This terminology of basic needs, rather than of rights or entitlements, displays significant continuities with the approach to water poverty in the years following privatization. Individual water companies responded to the plight of low-income families or those with special needs by creating water charities, to which those requiring exemptions or special treatment were required to prove their eligibility' (2001: 156).

In this context drought-sensitive behaviour management emerges and social media genres provide the key tools for combining reporting water with citizens reporting on one another's water use, wherein the 'public "shaming" of individuals and organizations considered to be wasting water is an increasing feature of drought events' such that

> Recent droughts have seen an escalation in the use of social media to shame excessive users and highlight leaks. In California, for example, it is common for residents to share pictures and videos of neighbours wasting water, using the 'Drought Shame app' and via other social media platforms. By November 2015, which was four years into the drought, state authorities encouraged the shaming process and launched a website (www.savewater.ca.gov) to which residents could send details and photos of water waste.
>
> (UNESCO 2016: 160)[25]

The shaming[26] on Twitter during the 2015 California drought was taken up by other countries through social mediation and gamification via moral and emotional stances to water scarcity. In Cape Town, South Africa, cinematic narratives of a countdown to Day Zero of total water shortage during 2018 produced a mediated narrative of 'the disaster movie-like moment when municipal engineers would have turned off the taps for millions and forced them to queue at military-guarded standpipes' states Joseph Cotterill in 'South Africa: How Cape Town beat the drought' such that:

> It is a powerful image of solidarity. The authorities have not been shy about reinforcing the message as they have kept the public informed of dam levels and water usage. A colour-coded map shows adherence to water restrictions street by street online – dark green for the virtuous, lighter shades for those under pressure to catch up. Escalating tariffs on the rich are meant to pay for basic water for the poor.
>
> (Cotterill 2018 [online])

This recreation of the modern media citizen as a good or bad ecological citizen is played out creatively through a convergence of mediations of 'drought shaming' (videos, GPS tracking, geo-tagged photos, stories, anecdotes and comments, memes, GIFs and soundbites). It positions individuals, whether they be members of the public or commercial and agricultural actors, land managers, or local and municipal governors as personally responsibilized through an always-on surveillance.[27] However,

---

[25] The authors of the report continue that: 'Although there is no clear evidence on the effectiveness of drought shaming in terms of water savings, it can clearly engage the public and raise awareness of drought issues' (UNESCO 2016: 160).

[26] See https://twitter.com/hashtag/droughtshaming?lang=en.

[27] For a more detailed sociological account of the drought shaming roles, performances, communicative acts and forms of media, as well as celebrity drought-shaming and tabloid coverage of the California drought, see Milbrandt (2017) 'Caught on camera, posted online: mediated moralities, visual politics and the case of urban 'drought-shaming' and Skoric (2010) on civic vigilantism.

before championing this as a new form of participatory and digital hydro-citizenship it does assume a proximity of the social media audience to the drought problem or that the social media activist has a pre-mediated sense of dryness in their own region that primes them to drought concerns.[28] It misses the key communication challenge around climate change and adaptation, that while a community or region has a pre-existing discourse of cultural identity built upon drought (e.g. Australian frontier spirit of greening the land), and thence dormant memories to be reactivated and remediated during drought events (see Anderson 2010), what of drought risk where there is *watery sense of place*? Endtar-Wada et al. (2009: 56) state that '[a]ttention to the "moral economy of water" recognizes that water is a "complex social good" that requires an "equally complex take on sustainability" centred on sustaining meanings that water has for people (Arnold 2009)' and that 'precedent and place' are key, but what if drought occurs where there is little precedent and no *dry sense of place*?

While UK consumers of water are being asked by the marketing and communications departments of water companies to be more 'water-wise' or 'water-smart', engage in 'water-saving' during 'water scarcity' and 'water stress', this makes clear that, discursively, the word 'drought' remains at a distance, that is, in the far past or far future or in historically dry countries such as South Africa, Australia or parts of the United States. As previous chapters have noted, there is a cultural and highly mediated perception that the UK is a small and wet island and that the infrequent recurrence of drought means that living memories of its impacts may be lost. In part, this is due to the readily accessible media templates of flood and deluge (see the previous chapter) and the hard to visualize drought impacts for a water risk that is slow, creeping and mostly hidden from view.

Yet it is clear that the flood-drought continuum needs better expression by all water stakeholders as well as national media representations and not left to the niche communicators of the scientific community or the agricultural sector, who may not get through to diverse publics.[29] They need to address the following local communication challenges that acknowledge the popular cultural identities framed by a watery sense of place: that Britain is wet and droughts simply do not happen; that if it does then drought only occurs in the summer, when it's hot; that drought is when rivers and reservoirs run dry; that rain spells the end of drought; that water is infinite and free

---

[28] For example, few Western media audiences knew of the 2006–11 drought problem in Syria prior to the outbreak of civil war, and rarely is the current war and political crisis (which has left so many dead, homeless and devastated) been represented as linked to drought. I learned of drought in Syria through a comic strip produced in 2014. See 'Syria's Climate Conflict' written by Audrey Quinn and illustrated by Jackie Roche, which has circulated through the Mother Jones website. Available at: https://www.motherjones.com/politics/2014/05/syria-climate-years-living-dangerously-symbolia/.

[29] While writing this chapter during the hot summer of 2018, my 'veg box' was delivered and with it the weekly 'News' from a UK farmer-entrepreneur who has created a successful but precarious business from British-grown vegetable farming and direct supply to consumers. On 9 July 2018 he wrote, the wet spring followed by lack of water have created the worst conditions: 'The only commercially viable option (and the most environmentally favourable) is to build clay-lined winter fill reservoirs wherever there is a valley bottom wide enough. To invest in such an asset that is used so unpredictably (on average every 30 years) is a bold move, but perhaps climate change is shifting the odds'. Two years later his business was struggling to 'feed the nation' during Covid19.

and can be readily accessed from Wales; that floods and droughts cannot occur at the same time; and, that droughts are short-term problems. Hence, drought dare not speak its name in a UK media context.

While all water companies in the UK must develop, and publish, a Drought Plan every five years (see the Water Act 2003) most of the companies do not 'front-end' their customer/consumer-facing communication with talk of 'drought'. During the UK DRY Project (2014–18) we found that sometimes when there is low river flow during water scarce periods, the water companies may communicate that they have plenty of water and the public should not be too concerned, thus placing customer relations management discourse ahead of broader behavioural change requirements to care for water. This can become further regionalized, with one water company expressing surplus while another enacts a drought plan, leaving 'customers' on one emotional side or the other, without a much wider sense of connectedness to national resource demands and limitations. Without creating negative communication, water companies frame water scarcity in eco-conscious and green discourses (asking consumers to care) with video and animated media that showcase technological solutions (meters and smart monitors) or emotional appeals to thinking of others and wider environmental concerns. For example, Thames Water's 'Be Water Smart' July 2018 promotional video begins by asking viewers to guess which city is the driest to counter the prevailing view that it is often raining in London. Partly, this maintains the discursive terrain of the UK as a wet place and drought as an aberration, with the consumer responsibilized for careless use of water, rather than the company publicly shamed for leakage due to poor maintenance, bad decision-making, lack of collaboration between privatized companies or any poor planning. What is striking in the tweets posted by Severn Trent Water on 28 June 2018 during fears of water shortage is that there is also a continued holding to account of the company by members of the public who show off their water smartness and attention to detail regarding the social media messaging.

> Reply 1: I hope @stwater are not suggesting that a little hot weather means we are short of water in Stourport-on-Severn. After all they are planning to extract a lot more water from the River Severn to supply Birmingham
> @stwater: Hi [Reply 1] *our reservoirs are looking healthy, we're just having to clean a lot more water at the moment due to the hot weather on Wednesday we cleaned 300 million litres as the demand has increased a lot more. Thanks* [. . .]
> Reply 2: Agree with the message, but 5000 litres is not 880 pints. It's just shy of 8803 pints. A pint is less that a litre, so clearly 5000 litres has to be more than 800 pints!

In response to this recognition of the water-smartness demonstrated by some UK Twitter users, I sought to establish a protocol during the DRY Project to address using the social web as a site of research alongside analysing the Project's own twitter feeds as circulators of drought communication.[30] Social media can help drought researchers in two key ways: to narrate the *liveness* or even the life of a drought story (during a

---

[30] The DRY project Twitter feeds had a main feed and one feed for each of the seven catchments. The About Drought project also had its own Twitter feed @Project_DRY.

drought event) either in the UK or abroad; and, to narrate the *liveliness* of a drought story (see Marres 2017: 107–11). It is important to note that these are two distinct modalities. The first is concerned with number (i.e. occurrence, currency, popularity and frequency) and means that as researchers we count plays, views, tweets, retweets, which are the simple repeats of the drought story with no modification or evidence of interaction. The second modality is concerned with media and cultural communication (i.e. issue-formation, composition, interactivity and user diversity/modification) and means we count retweets that modify by adding in new hashtags or @mentions, as well as allowing for the following risks and questions to be explored:

> *Babble* – does a story create unserious babble? Such as, a storyteller's friends view the video and have a laugh about a haircut or what they are wearing in the UK drought of 1976. Note, much science communications and professional social media influencing on drought issues is focused on issue-formation and serious debate of an issue. However, stories are also non-serious, engaging, fun, and we need to be attentive to the modalities of drought stories and users who engage in them in the ways they see fit rather than experts see fit.
>
> *Interface* – this is not a research method but a research interface, thus it opens our drought stories up to the potential of creative narratives, counter-narratives, connections and relationships
>
> *Backlash* – how does the approach above create room for objection, for counter story, for backlash or how does it elicit objections that in itself creates liveliness?
>
> *Profiles* – what hashtag profiles do other users create for our drought stories? How do they tweet about them and add hashtags and @mentions of their own choosing and how does that help us understand the narrative process?
>
> *Spin* – the 'spin' that others put on our drought stories in producing their own tweet 'orientates' the story in directions we do not anticipate. This creates conversation, debate and hydro-citizenship.
>
> *Bots* – some of the most engaging twitter interaction is created by 'bots' so non-humans, therefore, the role of the non-human network can be just as important as the human actor in water communication.

All of this would potentially allow us to see, through socially mediating drought story triggers, how the issue of drought is composed and recomposed by others.[31] If it goes nowhere and fails completely that tells researchers and social media communicators something very useful about drought stories in a particular context and network, the role of social media platforms in circulating them, and the vested interests of various stakeholders and actors in story-ing a scarcity of water in the stormy and immersive environment of online communication.

---

[31] Be mindful, when a colleague and I scraped twitter data on one day during a hot day in Spring in 2018 we found a high frequency of #drought topics related instead to a drought in goals during a Premier League football match, so quantitative data must always be checked for cultural codes.

Part Two

Culture

4

Story-ing water: Liquidity, bubbles, storage

Story-ing water

The previous chapter began to consider how everyday social mediations through networked platforms of media convergence *communicate* water as transmedial (see Harvey 2014)[1] and how citizens are co-producing water 'storyworlds' (see Ryan and Thon 2014: 1) of text, video, sound and emojis through everyday conversations online with water providers. Here, then, is some evidence that the *story-ing* of water is taking an increasingly media-enabled approach as part of a narrative turn. While macro-threats such as droughts in California that become global media events are obvious spectacles for social mediation and online emotion, it is the small and everyday encounter that is most illuminating, creating alternative stories, moving back and forth between scales of rights and responsibilities and demonstrating engagement with water issues at the quotidian level.

Such stories, anecdotes and snippets provide the fabula, the 'medium-independent' sense of 'raw material for textual elaboration' or 'a construct, abstracted from the text' as defined by narratology, wherein '"narrative" in this sense was "always already" (to use a Derridean turn-of-phrase) open to shaping in different media' (Rimmon-Kenan 2006: 13). Thus, if as Roland Barthes in *Image/Music/Text* (1977: 79) determined, narratives are a 'prodigious variety of genres, themselves distributed among different substances – as though any material were fit to receive man's stories' then what kinds of narratives absorb or saturate stories of water? Here, we may wish to make a distinction, as does Genette in *Narrative Discourse: An Essay on Method*, between story, narrative and narrating:

> To use the word *story* for the signified or narrative content (even if this content turns out, in a given case, to be low in dramatic intensity or fullness of incident), to use the word *narrative* for the signifier, statement, discourse or narrative text itself, and to use the word *narrating* for the producing narrative action and, by extension, the whole of the real or fictional situation in which the action takes place.

(1980: 27)

---

[1] Harvey draws on the fact audiences remember across media storyworlds. While he is focusing on fictional film and television franchises, we can use the same idea of a 'taxonomy' of transmedia to appreciate that water stories operate across media forms, genres and discourses and that how audiences determine what parts of a story of water to remember and forget can be connected to a sense of authorship and ownership of that story.

Thus, the water stories of the previous chapter, which often have very low dramatic intensity for the reader/audience (although not for the storyteller) place themselves, or are placed, in a overarching narrative of water as a receiving material. For Alleyne (2014: 2) 'narrative' refers to the presentation of a story while the story itself is the actual sequence of events; thus, he argues that '[n]arratives include stories but are not just stories: narratives are also about the way these stories are presented.'

From the moralizing on drought in shaming hashtags in Chapter 3 we can also see that environmental issues are highly emotionally charged and circulate as global stories of disaster and/or triumph (as explored in Garde-Hansen and Gorton's *Emotion Online* 2013) in relation to climate change. Obscured in and by these global texts and icons of extreme watery weather, introduced earlier in this book as 'media events', are the everyday water stories, lay knowledge bases, the local experts and the folk memories of water use and adaptation that often prove pivotal to exploring community resilience on the ground and across connected online communities.

From an arts and humanities perspective the reporting of water issues (as stories, anecdotes, snippets of narrative) from Chapter 3 aligns with the way media 'typically rely on narrative forms of communication, even when communicating important scientific information' (Jacobs 1996; Shanahan et al. 1999). Yet to assume these narratives have their intended impact is to imagine a rational subject of communication and reception in a straight channelled line of communication, when in reality *story-ing around water and water issues* is more like the experience. Thus, I wish to move us beyond metaphors of media as flow, channels, immersion and streams in the analogue-to-digital age and the uncontrollable storms and tsunamis in the social media age, to some different structures? If these conceptualizations of media, culture and communication suggest we are building our constructs upon an abundance of water and weather to be engineered, controlled and culverted, then how is this resilient? Where are the metaphors of pollution, drought, scarcity, trickle and shortage in our new watery mediascapes, and in our story-ing of water? If, as MacIntyre (2007: 216) writes 'man is in his actions and practice, as well as in his fictions, essentially a story-telling animal' then the question is less about what one should do about water issues, and more about what kinds of water stories do we find ourselves already a part of?

Consequently, unlike a social sciences approach that may seek to map, tag, order and node these stories, perhaps to place them in pre-existing or emergent narrative themes, as Andrews et al. (2013: 13) view 'ordered representations [that] can indeed claim to be mapping forms of *local* knowledge', we should realize that media make mapping messy. Media also breathe a certain afterlife, ghostliness and spectrality into stories, narratives and narrating, allowing them, however momentarily, to recirculate, drift, float and expand:[2] much like the bubbles blown by the child in Peter Sloterdijk's *Bubbles: Spheres 1 Microspherology* (2011) in which he begins his philosophy of spheres which we are always inside, with the following observation:

---

[2] Of television Joe Moran says 'for a programme that millions once watched but which has now faded into the atmosphere like a dream is a neat encapsulation of the elusive quality of memory itself' (2013: 5–6).

While exhaled air usually vanishes without a trace, the breath encased in these orbs is granted a momentary afterlife. While the bubbles move through space, their creator is truly outside of himself – with them and in them. In the orbs, his exhaled air has separated from him and is now preserved and carried further; at the same time, the child is transported away from itself by losing itself in the breathless co-flight of its attention through the animated space.

(Sloterdijk 2011: 18)

His 'spherology' inspires his approach to media studies. Media create prosthetic bubbles for anxious humans to be inside (eg protected from greenhouse gas spheres). In this chapter of Part Two on Culture, I wish to move the reader away from engineered metaphorizations of the relationship between media and water (channel, flow, contra-flow all of which assume abundance and plenitude) towards a more diffuse, de-territorialized and de-colonizing narrative framework for water stories, perhaps informed by postcolonialism but mostly from my water memory research in Brazil. As Pandurang puts it:

Current analytical models are not adequate for exploring new forms of multi-culturality that are in the process of emerging. What is needed is a theoretical framework that goes beyond formulations of cultural imperialism and simplified binaries and speaks from the affective experience of social marginality and from the perspective of the edge – they offer alternative views of seeing and thinking, and thereby allow for narratives of plurality, fluidity, and always emergent becoming.

(2001:2)

Therefore, the circling around of water stories and water issues and the new attention to storytelling in the science community sets the scene for future chapters which consider the multimodality, intermediality[3] and media convergence at work in digital hydro-citizenship that affords new forms of storytelling. For now, we shift from a focus on communication through media to media in culture. This is to open up the story-ing of water to multiple modes and multiple directions, and to consider what narrative structures water stories are keyed into, as well as how water is becoming popularized, mainstream and an entry-point story for wider concerns.

Water is *on trend* right now

According to Dahlstrom (2014) there are four parts of a narrative that improve knowledge acquisition: dramatization, emotionalization, personalization and fictionalization. All these parts of a narrative are at work in the story-ing of water from science, to fiction, television to film, news to politics. During the writing of this chapter water issues were high on the national and international agenda and

---

[3] Ryan and Thon (2014: 9–10) define '*multimodality*' as 'different types of signs [that] combine within the same media object – for example, moving image, spoken language, music, and sometimes text in film – while through *intermediality*, texts of a given medium send tendrils toward other media'.

circulating widely within popular culture.[4] The United Nations Framework on Climate Change (COP24) panel met in Katowice, Poland, in December 2018, and Sir David Attenborough, the BBC's most popular natural history presenter was taking up the People's Seat, and warning that time was running out for decision-making on climate change scenarios. Attenborough's acclaimed and widely popular *The Blue Planet* and *Blue Planet II* (BBC 2001/2 and 2017/18)[5] series have had a significant impact on the mediated collective consciousness of water (oceans, seas, coral reefs, rivers, etc.) and the pollution, particularly plastic pollution,[6] which is increasing in those waters. Yet there is an argument to be had about the underlying economics of this documentary series' spectacular success and the kinds of viewers it engages and the calls to action that may or may not ensue in response. Wheatley in *Spectacular Television: Exploring Televisual Pleasure* (2016) makes the important point that *The Blue Planet* (2001), which was actually co-financed with the Discovery Channel, and broadcast the day after the horrific media spectacle of 9/11, was perfectly designed for the Bang and Olufsen viewing culture, with a 'technologically enhanced form of viewing in mind' such that 'the audio-visual pleasures of *The Blue Planet*' were 'best enjoyed by those with capital (both cultural and actual) to appreciate them' (Wheatley 2016: 97–8). Story-ing water in *The Blue Planet* (BBC 2001) is much less about narrative, claims Wheatley, and far more about stretching genre.[7]

With far 'greater emphasis on abstract visual and aural pleasure' than in 'earlier examples of the genre', Wheatley supports her ideas with quantitative comparative analysis of *The Blue Planet* with the earlier *Life on Earth* (BBC 1979) series also narrated by Attenborough, and qualitative interviews with audience members. Wheatley's

---

[4] For example, on 20 December 2018 the Wall Street Trader Victor Vescovo became the first person to reach the deepest point of the Atlantic Ocean, the Puerto Rico Trench. The year saw *The Shape of Water* (directed by Del Toro 2017) win an Oscar, *The Water Cure* (2018) by Sophie Mackintosh received critical acclaim as a debut novel and popular science books on water proliferated such as *Tides and the Ocean: Water's Movement around the World, from Waves to Whirlpools* by William Thompson (2017).

[5] The use of highly skilled production crews filming in hundreds of locations and at sea depths hitherto unseen by film cameras, at scales not possible in previous nature documentaries, positions both series as 'event TV' or 'appointment TV'. Attenborough is widely considered a 'national treasure' and voice of the people, and the series has produced numerous paratexts (merchandise, books, a live concert tour, etc.) and has sold into fifty different countries.

[6] The Royal Statistical Society ran its 'statistic of the year' on 18 December 2018 and the 'winning International Statistic of 2018 [was] 90.5 per cent: the proportion of plastic waste that has never been recycled. Estimated at 6,300 million metric tonnes, it's thought that around 12 per cent of all plastic waste has been incinerated, with roughly 79 per cent accumulating in either landfill or the natural environment (as sourced from "Production, Use, and Fate of all plastics ever made" by R. Geyer, J. R. Jambeck and K. Lavender Law 2017)'. See website https://www.statslife.org.uk/news/4026-statistics-of-the-year-2018-winners-announced.

[7] The genre has expanded. For example, of Christmas television in the UK between the usual festive films and dramas, we have repeats of *Frozen Planet* (BBC 2011–12), *The Blue Planet* (BBC 2001) and a genre of television I term 'weatherized', which generally comprises the annual taking stock of the year's weather: *Spy in the Snow* (BBC 2018), *Britain's Wildest Weather* (Channel 4 2014–2018) and *Wild Weather UK: Winners and Losers 2018* (ITV 2018). In covering 2018's weather the Sky News website read: 'From droughts to deluges: A year of dangerous weather' at https://news.sky.com/story/extreme-weather-2018-snow-in-the-sahara-wildfires-and-typhoons-11471584. Accessed 31 December 2018.

hypothesis is that 'the *beauty* of what was being shown', particularly the underwater footage clipped for advertising and other paratexts (the latter term explained in the next section) superseded both narrative voice over and story 'in favour of a soaring orchestral score' (Wheatley 2016: 98).

In measuring the ratio of narrated to non-narrated footage in a selection of episodes from both series, Wheatley notes that *The Blue Planet* had less narrated sequences, and unlike *Life on Earth*, Attenborough's voice-over was an in-studio post-production paste-on. This is not to do a disservice to the high-quality on-location and studio production; rather, Wheatley emphasizes the mode of address in order to tease out the intended audience (clearly with cultural taste) and the emotional effects (spectacular, memorable, haunting). *The Blue Planet* was 'beautiful television', a justification for paying a licence fee, as well as a visual feast (Wheatley 2016: 100), presenting water as precious, mysterious and breathtaking but with a nod to a discerning audience of higher cultural tastes. Since *Blue Planet II* (BBC 2017–18), an equally spectacular documentary but with the added dimension of a socially mediating audience, we have extensions of the spectacle into Live Events[8] as well as global events such as the launch of the first Netflix natural history series *Our Planet* (2019) again using the Attenborough formula but with a much clearer climate change message to a global television audience.[9] It is unsurprising, then, that in my local bookshop, while *Blue Planet II Live* was being marketed, that the bookseller was curating established and new books for the Christmas gift-giver, that signalled a watery and maritime turn in arts, culture and humanities as noted in the Introduction, but here for the general reader and water enthusiast (see Figure 4.1).

Similarly, in August 2018, Mary Beth Griggs noted in an editorial for *Popular Science* magazine in the United States that the Strand Book store in New York City had curated eighteen water-themed books with the *Popular Science* team for a stand entitled *All Wet*:

> This month we're featuring books about a substance that defines life on Earth more than any other: water. Did you know that 71 per cent of the Earth's surface is covered by water? Whether you look at the glass as half-full or half-empty, salty or fresh, home to sea creatures, or frontier for exploration, there's something for everyone.
>
> (Griggs 2018 [online])

Why, then, this upsurge of water stories now and why is it worth noting for media and cultural studies scholars? I think for two main reasons, the first is how water is being used to articulate the storytelling process in an age of uncertain futures. Rather than

---

8   *Blue Planet II* has extended its paratextual presence for the discerning audience of cultural taste by touring a Live in Concert during 2019 with 4-K-Ultra HD LED Screen alongside the City of Prague Philharmonic Orchestra to over six major cities. The half-page advertisement in *The Guardian* newspaper on 22 December 2018 described it as 'the perfect Christmas gift'.

9   *Our Planet* (Netflix 2019) has three of the eight episodes focused on water: 'Coastal Seas', 'The High Seas' and 'Fresh Water'.

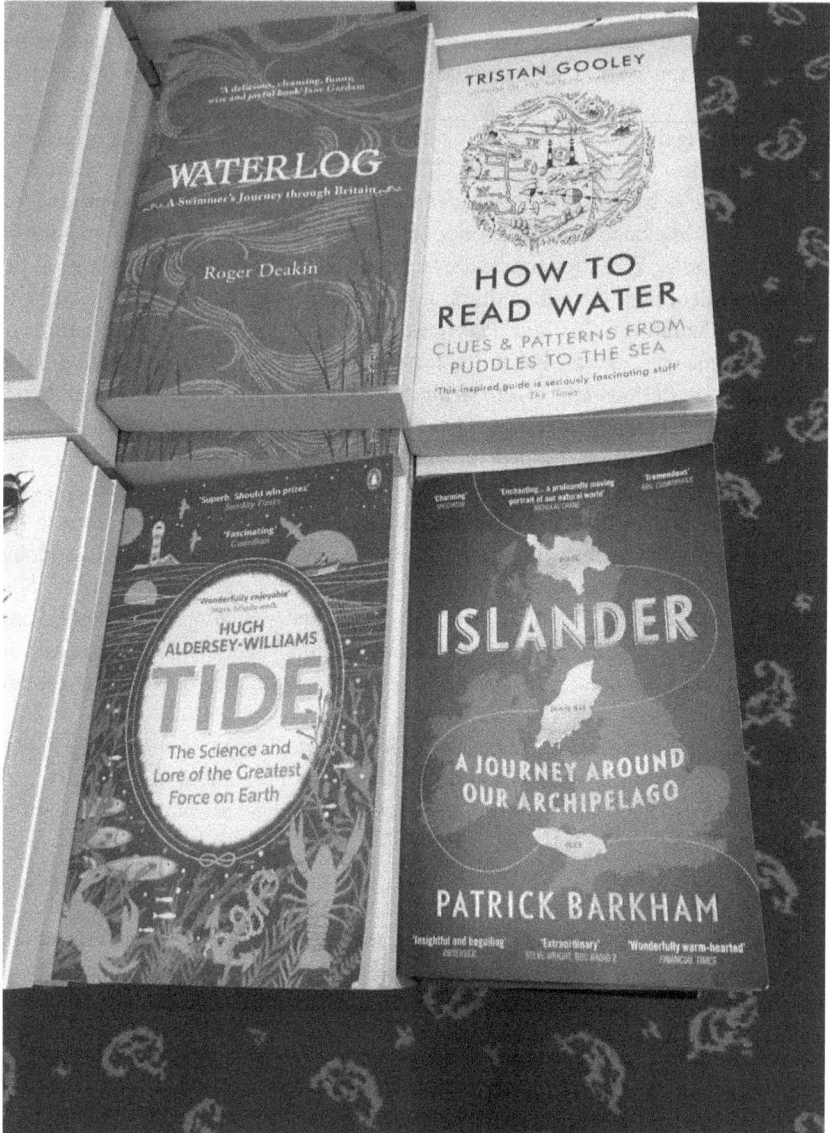

Figure 4.1 Bookshop display, Leamington Spa, December 2018.

channelling the narrative in one prescribed and engineered direction, as one did in a broadcast landscape, storytellers are drawing on water for its immersive, meandering and submersive capacities: 'A river no more begins at its source than a story begins on the first page' so says Diana Setterfield in her fictional novel *Once upon a River* (2018: 27). The second is that the water story has become the paratext of the climate change narrative text, and it is worth dwelling on the term paratext within media and cultural studies for considering the new liquidity (in commercial and social terms) of stories of water.

Paratextual liquidity

Genette in his 1987 book *Seuils* translated into English in 1997 as *Paratexts: Thresholds of Interpretation* argued that the paratext is the term

> to describe all those things that surround the actual literary work that we may be inclined to consider not wholly a part of it, but that nevertheless append themselves to it, whether physically, as with book covers, prefaces, afterwords, and choices over paperstock and typeface, or conceptually, as with reviews, interviews, ads, and promotional materials.
>
> (Brookey and Gray 2017: 102)

If water is becoming increasingly mediated and mediatized, if water has been intensely storied and story-fied in recent years, in such a myriad of modes, genres, spectacles and styles, then what is the main text, the main narrative, the big story? I would argue that stories of water function as keys or entry points into a much larger climate change discourse and narrative, where national and global concerns are in conflict. Climate change (at least, the representation of it across a range of genres from the scientific to the dramatic) is the main text, the actual work, the big one; and, it is the entity very firmly embedded in society and culture, whether it is believed or denied. Intrinsically part of the narrative of climate change and yet running in all sorts of directions within, through, alongside, foregrounding and backgrounding, is the water story, the paratext. Since Genette's formulation (obviously rooted in literary studies) media and communication scholars have begun 'applying the term to film, television, and other media' because it covers the story-world of the original text, the 'huge world of promos, hype, trailers, merchandise, licensed games, DVD bonus materials, ancillaries, transmedia extensions, fan texts, and more' (Brookey and Gray 2017: 102). I would extend the paratextual to encompass all forms of cultural and media representation that create a following, an audience, a target group or social and civic calls to action, even if those narratives are concerned with the communication of science. Since 'we regularly do "judge a book by its cover," Genette argued that textual analysis must account for the meanings that covers and their paratextual colleagues create' say Brookey and Gray (2017: 102); and, if this is the case, how do we judge climate change through the stories of water?

Stories are knowledge in action (see Bacallao-Pino 2016). Anecdotes demand to be told says Mike Michael (2012). Shared through formal networks (such as journalism) and informal networks (such as families), stories of water can be studied for their potential to circulate, actuate and activate a response to water management issues. From God's commandment to Noah to the Babylonian epic Utnapishtim, from Deukalion to Atlantis, knowledge of extreme water events has been inherited as oral stories of strategic response to risk, often in which a male hero triumphs by following a command and control policy or as trauma and devastation. Thus, while there is a great deal of literature on extreme weather and water events (as noted in previous chapters), and while the evidence upon which this is largely couched in the natural and social sciences, it is becoming increasingly apparent that water stories (literature, theatre, art and media) and the story-ing of water (from the professional to the personal) have connectivity across times, places and peoples, and exclude as much as they include. Water stories may also serve to enshrine knowledge and action in conventional power structures and in particular locations for certain beneficiaries and stakeholders.

If stories are knowledge in 'action', this begs the question, is the expected action in the hands of an expert few who have the power of omniscience, prediction and forecasting or is it also necessary to offer old, new and counter-narratives to those dominant accounts which from centres of expertise or 'centres of calculation' as Bruno Latour (1988) defines them? If so, what should be the role of the masses, the popular, the 'common culture', the mass media and increasingly social mediators in distributing stories more widely? Importantly, how are authoritative knowledge and a command and control approach to water management being addressed by activists, social movements and the direct actions of concerned citizens? In what follows, I wish to set the scene on water story-ing and stories of water in terms of the liquidity of their paratextual status, as emotional, visual, spectacular keys to larger, grander narratives that circulate and create value and values. The literature cited in this chapter is set to frame the following two chapters which consider more deeply case studies on drought histories and gender, forgetting floods and the use of a flood memory app to collect snippets of stories and geotag them to broader stories of living with water. In all the cases, water stories become keys to broader narratives of risk, resilience, threat, fear and as the late cultural critic Mark Fisher so eloquently put it 'the slow cancellation of the future', which 'captures very well the sense of the ebbing away of a certain conception of cultural time' (Fisher and Berardi 2013 [online]).

Such changes in cultural perceptions of time are addressed in *Emotion Online* (2013), in which Garde-Hansen and Gorton textually analysed the online debates around some climate change media that entangled the threat of bad weather with fears over race relations, war and political dissent. Drawing upon Massumi's (2011) work which connects extreme weather and war, we proposed the concept of an '*emo-techno-ecology*' to address the way these changing-environment-fears exist 'trans-medially' but are felt intimately at 'hyper-local' levels. Thus, we argued that 'we need to understand our mediated ecology along two wavelengths simultaneously: as local and global emotions' or as 'global emo-scapes' in which citizens are affectively connected to their environment as 'technologically enabled infotainment producers/consumers' (Garde-Hansen and Gorton 2013: 128) but are also called to action on the ground. More broadly, Brace and Geoghegan (2011), writing in the context of human geography, argue that climate change is encountered holistically, not just in how it is understood 'top-down' through the communication of scientific discourses but also relationally at a local level:

> Climate change can be observed in relation to landscape but also felt, sensed, apprehended emotionally as part of the fabric of everyday life in which acceptance, denial, resignation and action co-exist as personal and social responses to the local manifestations of a global problem.
>
> (Brace and Geoghegan 2011: 284)

Thus, there is a new liquidity at work, a new circulation of values of water, that even, when technologically enabled, allow citizens to *bank water's cultures,* to invest in watery terrains, to purchase more books on water, to vicariously swim where others have dared to swim, and to emotionally connect themselves to water issues. This is why place, territory, terrain and occupying watery spaces (literally and imaginatively) are so crucial to water science and water management.[10] However, before we consider further the storage and banking of water stories and water cultures, let us understand media as bubbles.

Story bubbles

If there are communication barriers to representing or shaping water policy, practice, management, governance and security then these can be made transparent by drawing into our approach to water storying Sloterdjik's concept of bubbles from his 'sphere theory' (cited earlier) to counter the idea that there is a straight channelled through line from communicator to target audience that has the most impact. In their recent article 'Narratives as tools for influencing policy change' Crow and Jones (2018) take the classic marketing and communications position that all one needs to focus on are the following: setting, characters, plot and the moral of the story (which is, for them, the call to action). This assumes that senders of messages (policymakers) and receivers (the public) are not creative and critical storytellers in their own mediated spheres. Thus, it may overlook the idea that both sides of this communication are asking: 'Whose story is it? Who authored the tale? Whose voices were included? Whose voices were silenced? As our attention is called to one facet or event, what aspects are nudged into shadow?' (Morgan-Fleming et al. 2007: 82). Addressing these kinds of questions as they play out in stories of water allows us to inhabit 'two incommensurable theories of the speaking subject: the agentic, storytelling subject of the experience-focused tradition, at odds with the fragmented disunited "postmodern" subject of more culturally-oriented analyses, produced by the cultural stories around them' (Squire 2008: 48). These subjective spheres exist side by side and media become the bubbles that are the 'messages, senders, channels, languages' that are 'the basic concepts [of media theory], frequently misunderstood, of a general science of visitability of something by something in something' (Sloterdijk 2011: 31). Thus, '[w]e will show' says Sloterdijk 'that media theory and sphere theory converge [...]. In spheres, shared inspirations become the reason for the possibility of humans existing together in communes and peoples' (2011: 31). While bubbles have received a bad reputation of late, particularly 'filter bubbles'[11] (the metaphorization of consumers of news, for example, that suggests opinions float unchallenged and unpunctured in a soft

---

[10] During the writing of this chapter, the author was successful in receiving a Coventry-Warwick City of Culture funding award for a project entitled: *To Walk with Water: Young People and Urban River Cultural Values in Spon End, Coventry.* This was a collaboration with the Centre for Agroecology, Water and Resilience, Coventry University and Talking Birds, a nationally renowned company of artists specializing in theatre of place. During the project we were able to collect the stories of marginalized water users on the margins of the River Sherbourne.

[11] See the widely viewed online talks of the American internet activist Eli Pariser.

and protective sphere of homogeneity) this misses a suggestion that being in a sphere is a key part of story-ing the self.

> What recent philosophers referred to as 'being-in-the-world' first of all, and in most cases, means being-in-spheres. If humans are there, it is initially in spaces that have opened for them, humans have given them form, content, extension and relative duration. As spheres are the original product of human coexistence, however – something of which no theory of work has ever taken notice – these atmospheric-symbolic spaces for humans are dependent on constant renewal.
>
> (Sloterdjik 2011: 46)

While they do not describe it in terms of a water story sphere, Bryant and Garnham (2013) expand on the same kind of 'being in the world' in an Australian context of drought risk, citing many scholars, in their explanation of life predicated within, upon and beside water, as:

> intimate and complex[12] intersections between home, business and family contextualized upon the site, or place, of the 'farmland', this formation constitutes more than a business or occupation to become a 'way of life' (Price and Evans 2005, 2009; Ramirez-Ferrero 2005). This way of life is characterised by cultural meanings and values relating to conceptions of rural idyll (Bryant and Pini 2011), gendered identities within a patriarchal and often patrilineal system of farming (Alston 1995; Bryant 1999; Panelli 2002) and a macro political-economic context (Lockie et al. 2006).
>
> (Bryant and Garnham 2013: 3)

A sphere approach to story-ing water may then be necessary for understanding the '"tiny affective nuggets" that circulate and accrue' in mediated spheres where care, compassion and responsibility are at the forefront of the discussions (Wilson and Chivers Yochim 2017: 27–8).[13] If water stories are in circulation as such 'tiny affective nuggets', so too must research communities who follow water issues be attentive to filtering of stories into climate change or denial narratives that may lose these small stories of affective resistance to what appear as obvious water adaptations.[14] One perplexing micro-narrative to agencies is the refusal of certain householders to

---

[12] A good example of complexity might be in Head, Lesley M. and Pat Muir (2007) who found that a commitment to reducing water consumption was shared amongst Australian garden owners, but they concluded that people's water aspirations were in tension with pleasure derived from water and with a desire for more watery environments.

[13] On women's use of online forums for discussion, it is worth noting that Wilson and Chivers Yochim are defining the 'mamasphere' as a 'contradictory web of advice, friendship, information, and entertainment, fuelled by highly-organised and interactive data-mining machines, but also by the situated experiences of mothers' (2017: 27).

[14] Chappels et al. studied the 2006 drought in South East England for the impact it had on outdoor domestic water consumption habits. Responses to hosepipe bans were variable which led the authors to conclude that the extent of change is mediated by 'pre-existing social orientations and by diverse configurations of garden infrastructures and water institutions' (2011: 701).

move home despite repeated inundation, or evacuate, which is not easily explained in economic terms. What kinds of affective and social power and solidarity do this resistance to flood water engender in the creation of a coping community? What kinds of normative gender identities resist drought realities in the face of economic and mental stress (see Alston and Kent 2008)?

In my own research, I have captured some of the small stories of water through a variety of media, art, literature, music, memory work and photography and in the collaborations I have undertaken with key water researchers in the UK (see Figure 4.2 for how these bubbles overlap and connect to one another), I have found a sphere approach to investigating water the most productive and much of the evidence from these research projects informs the rest of the chapters in this book. Starting with the contribution I made to the Living Flood Histories project, in which I proposed in 2011 the development of a flood memory app (see Chapter 6) to collect affective nuggets of water stories, this led directly to the Sustainable Flood Memories project based upon the 2007 Floods (when Twitter and Facebook were still finding their feet but local news organizations, radio and television mediated the memories). A knowledge exchange project with the Environment Agency extended the digital storytelling of flooding in the UK by sharing DIY narratives across river basins/catchments with their own 'watery sense of place'.

In our research for the Sustainable Flood Memories project we discovered that to support remembering, many spheres had to bubble together: homes, gardens, streets, businesses, churches, riverbanks, urban infrastructure and artefacts were mediated using photography, home video, broadcast media, social media and, overall, the internet

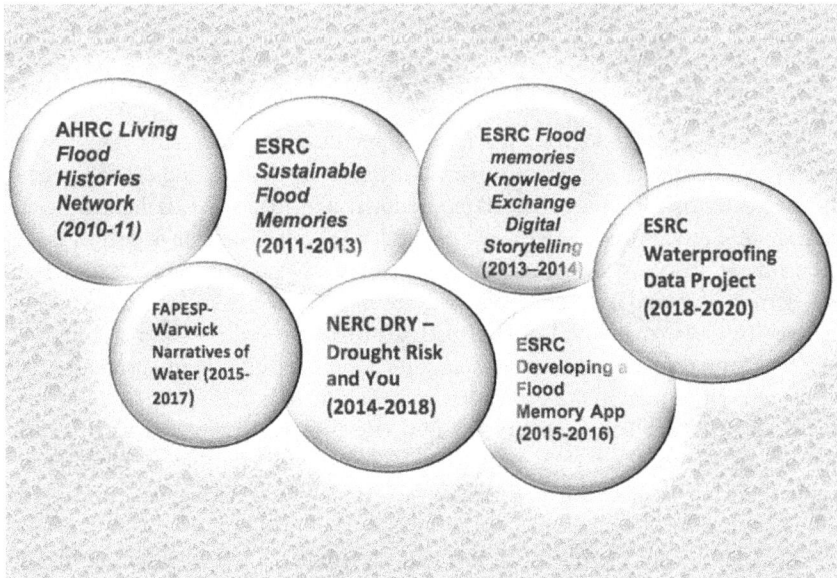

Figure 4.2 Image of Bubbles of Water Research: Most of these research projects. I have either led, co-led or (in the case of the Living Flood Histories Network) contributed to.

into a social and technological *memory bank* of stored flood knowledge that could be mobilized by journalists, citizens, communities and archives. In terms of forgetting, we also found these same spaces adept at erasing a watery sense of place (changes to street names, removal of flood marks, flood image fatigue and loss of archival images and footage). Spheres of memory and amnesia bubbling around one another. Thus, one outcome of our research was to understand the role of forgetting: defined as repression, amnesia, annulment, erasure and planned obsolescence by Connerton (2008) as significant spheres of human habitation in which some communities seem to float above flood-risk narratives, working against a sustainable flood memory, which we have written about elsewhere (see McEwen et al. 2016; Garde-Hansen et al. 2016).

The flood-drought continuum, wherein two contradictory spheres coalesce and touch one another, led to a much larger project on narratives of water scarcity, the Drought Risk and You (DRY) project (2014–18) and this is connected to the other water story research projects in the UK taking place at the same time and connected through collaborating researchers: towards Hydro-citizenship (2014–18) and Multi-Story Water (2012–17). The more applied research on developing a Flood Memory App alongside the strand of transnational and transmedia research with Brazil on Narratives of Water (2017–18) and Waterproofing Data (2018–20) are part of a programme to connect transnational and transcultural spheres through digitally mediated flood memories and stories. Much of the research mentioned above suggests that for a country surrounded by water, where talking about the weather is not only an everyday past time but key to cultural identity, the UK is producing spheres of multi-disciplinary water research that are increasingly connected to one another and yet will remain in their own spheres if media and storytelling are not taken more seriously.

Storage

As noted above, and underpinning forthcoming chapters, my media research has explored water and narrative to understand the mobility of flood and drought media, memory and storying. I have focused on the flows and frictions across different socio-demographic groups who live with a watery or dry sense of place, alongside the value of the memories to be abstracted from these environments as a future resource for action. The starting point and end point have always been the person as 'an intangible cultural heritage' (see Worcman and Garde-Hansen 2016), as a storyteller with an increased capacity to use media as a social technology for communicating and remembering. As I have recently written in the context of media heritage: 'In a landscape transformed by online media and culture; interactive film, television and radio; and faster, mobile internet, the viewer formerly known as the audience will want the most memorable aspects of their mediated lives available, accessible and possessable' (Gorton and Garde-Hansen 2019: 42).

The ability, then, of this story and narrative research of water to inform our understanding of and respect for water and its movement through built, natural, canalized, neglected and polluted landscapes has been the aim but how to ensure these water stories are stored? To fully understand the significance that climate change impacts will have, we will need conditions that require and engender remembering. Not only that, we will require cultural work for Allen argues

> Remembrance implies work. It requires bodies and objects moving, performing and interacting to achieve a memorial composition. Participants in remembrance practices undertake a material and immaterial work of weaving their emergent experiences as part of the overall composition of the event. A material management of space, objects and bodies structures experiences of participation. The organisers of remembrance practices, memory choreographers, are responsible for designing and assembling these infrastructures of experience.
>
> (Allen 2014: 28)

Like memories of media and cultural productions, memories of water stories are monumentalized, perhaps not always in stone and bronze, though we do have plaques, commemorative events, museum exhibitions and archival holdings to remember flooding.[15] There are physical manifestations of water inheritance (engineering works, museum artefacts, water technologies and collections of memorabilia for maritime, fishing, leisure and agriculture). Some of these water memories punctuate the landscape as flood marks, restored canal boats, old river walls and bridges, maritime heritage and statues of seafarers. Most water memories are intangible.

In the urban environment, for example, floods can be catastrophic. Flood risk and flood events will require a rational multi-stakeholder response, debated in the public sphere. Yet the stories, memories and images of flood risk are loaded with aesthetic, emotive and performative power, which has memorability.[16] More pertinently for media studies and for those researchers interested in the mediation of memory, as nature penetrates the domestic scene, it is the transformation of the 'home video' into evidence (such as in Figure 4.3) for storage by the homeowner that offers the intimate sphere of mediated water permeated by the public sphere of shock, debate, opinions and calls to action.

Here, media offer many bubbles, as citizen witnessing of environmental damage, of family mementos for future generations, of stories of active helplessness, and more importantly for many of the participants in our research, as visual proof to the insurance company that one's private sphere was now connected to water risk and stories of water damage. Mediations of flood as family stories also reveal the underlying

---

[15] For example, the Houston Flood Museum documenting the impact of Hurricane Harvey (2017) where you can tell your story and submit it to the online 'museum' with images, providing the location and narrative data. Accessible at: https://houstonfloodmuseum.org/.

[16] There are numerous home videos of children and teenagers enjoying and playing in flood water on YouTube, as well as footage of people becoming illuminated by their sense of a community coming together and having a role to play. I find these particularly memorable for their counter-narrative qualities, and in our *Sustainable Flood Memories* project older residents would frequently recall with happiness being allowed to play in flood water when they were children. Such a re-framing of flood as opportunity is now permeating academic discussions, see 'Managing flooding: From a problem to an opportunity' by Richard Ashley et al. (2020), and signals flood as more consistently present in daily life.

Figure 4.3 Still of home video of River Severn rushing through a resident's home during the 2007 Summer Floods. Image courtesy of the resident who took part in the SFM project.

economies of flood memory. For there is a close, complex and non-linear relationship between memory, heritage and identity, with a fundamental dichotomy in ways in which heritage can be understood (see Anheier and Raj Isar 2011). In our research on flood media and memory we uncovered a hidden lived memory that demonstrated an environmental literacy towards resilience on the one hand and a precarity towards the storage of memories on the other. Flood knowledge that is held by a particular generation may not be passed down as a form of water heritage, and this challenged the notion of heritage as an essentially conservative, nostalgic project which encompasses a romanticized and idealized view of the past, reinforcing old certainties at times of significant change (Lowenthal 1985, 1988; Hewison 1998). It also provided evidence for a view that recognizes a more democratic form of heritage, emphasizing the 'spirit of local places' and the 'little platoons' rather than the 'great society' (Samuel 1994: 158).

Formal and informal flood marks on buildings and homes are a water story storage method. Not only through official and unofficial flood markers, placed on an abbey wall (one of many markers over the centuries) as a reminder to all who visit, but also on the farmhouse wall as a family reminder to future generations. Built walls, embankments, flood defence systems, an extra course of bricks, sandbags, raised doorsteps, electrical sockets at waist height materialize an environmental literacy that seeks to 'remind' us that the river has and will make its mark in these human spaces. Water memories are story-ed in multiple and connected spheres that media can network. Personal and collective memories of flooding and flood risk from media and archives play a significant and often under-researched role in the individual, community or society's sense of preparedness, and anecdotal memories demonstrate traditions of storytelling, informal remembrance inherited within families and the challenges of persistence and transiency of materialization in an increasingly digital society. In a community defined by its water heritage, a dynamic interplay of active remembering and active forgetting are at work. Hence, it is not memorialization that is celebrated but 'memorialism' (Dicks 2000), which is an attempt by local communities to make and maintain their own heritage.[17]

All this accords with research we published in 2012 when we produced a volume of chapters on bringing geography and memory studies together for the first time. 'Given memory's fundamental role in individual, family and other small collectives, and the always-present aspects of space, landscape and place within the memories of such (and vice versa), there is a sense that these "smaller scale" dynamics of geography and memory remain under-represented and less-considered within both memory studies, geography and other disciplines' (Jones and Garde-Hansen 2012: 3). Here, we drew together the new conceptual and empirical emphases upon performative and embodied practices of everyday life through our attention to the work of Nigel Thrift (1997, 2004). Thrift's (1992) early ideas on globalization were developed into an application of affect within cultural geography that has been important for resituating global concerns as not simply abstractly social-constructionist. Rather, 'care, risk, fear, responsibility, contentment, self-control, anger, shame, desire and hate' come to reemphasize 'affective personhood' to address 'local, national and global calls for individual and connected practices that are creative, sustainable, open, shared' (Jones and Garde-Hansen 2012: 5). Thus, *emotional geographies* should be understood as, first and foremost, the affective dimensions of home, space, place, landscape, area, environment and atmosphere as a priori experiences that move us, and that we move in and through. They are mobile emotional geographies. However, while traditional broadcast media may remember these emotional responses through human interest stories fixed in time and space, individuals in their own communities and contexts record their experience of extreme weather, water events and water issues on a personal level and connect those stories and memories on a range of scales, regionally, nationally and globally. They keep these memories mobile and anecdotal (that is, as anecdotes that demand to be told) and increasingly networked. The call to action is then as much emotional, personal and embodied (which is often forgotten) as rational and thoughtful.[18]

---

[17] See 'Heritage from Below' (Robertson 2008; 2012) and Laura Jane Smith's work and her identification of the near-universally accepted Authorised Heritage Discourse (AHD). In stressing the fluidity of both heritage and identity, Smith argues strongly for the understanding of heritage as 'something vital and alive ... a moment of action' (2006: 83). Smith's thesis, therefore, understands heritage as active and processual.

[18] Bear and Eden (2011) examine how freshwater anglers in northern England 'read' rivers, working with them relationally through various embodied knowledge-practices known as 'watercraft'.

There are collective memories of water mediatized by national and international storytellers (e.g. news agencies, films, radio broadcasts, literature, documentary photography), and these can be said to accord with the sociologist Halbwachs' (1992) conceptual framework of collective memory, as requiring the support of a group, wherein the totality of the event can be put into a single record. They may also, as he suggests, come into being through a separation from the lived experience of the community that experiences their watery or dry sense of place.[19] Moreover the increasing connectedness of collectives through digital media finds water stories travelling in new and different directions and at interrelated scales of personal, local, national and global.

The culmination of this research into story-ing water has fundamentally changed and challenged my approach to both media and memory studies. The circulation of water issues in a global narrative of climate change, the cultural niches connected across national boundaries of riverine communities and the shared stories of water (its abundance and scarcity) have produced connectivities with media history, media heritage, cultural value and cultural policy that disrupt the neat distinctions between science and narrative, data and story, non-human and human. Thus, mediations of water stories draw in articulations of water's chemical profile and the river's personhood and open up story-ing water to many new questions. Does the river have a memory? Are its chemical constituents an archive of human histories? How does water contribute to the social mentality of the communities that dwell, remember, forget, neglect or nurture it? Does the river always remember, in the end, where it once meandered? If the river is designated as a legal person (as the indigenous river and the new articulations of water in New Zealand have shown[20]) then what agency does it have? What are its rights and responsibilities towards us and us towards it; and who is this 'us'? What can we learn from water that has been nurtured or neglected, remembered or forgotten? When we view the city from the urban river's perspective, what do we see? When we view the flood water with an underwater news camera, how does it make us feel?

We may be at a watershed moment. Not only in river chemistry wherein the health of the river reflects land use and human behaviour but also in the storytelling of water and in our attention to cultural and collective memories of water. Too much focus on scientific instrumentalism to water scarcity, demand and management places economic values on water at the expense of cultural and social values. Allon and Soufolis (2006) make a similar argument that the environmental instrumentalism of current resource-centred approaches in Australia that ignore 'social and cultural differences associated with different habits, expectations, meaning and practices of water use' (2006: 46). The point here is that what we understand and produce as 'the river', 'the sea' or 'water' differs across cultures, where what is a life affirming resource

---

[19] Sunil Amrith (2020) has reimagined South Asian history in *Unruly Waters: How Mountain Rivers and Monsoons Have Shaped South Asia's History* by drawing on fears and dreams of water.

[20] As the New Zealand government website states: 'In 2019, New Zealand will mark 250 years since the first meetings between Māori and Europeans during James Cook and the Endeavour's 1769 voyage to Aotearoa New Zealand'. Available at: https://mch.govt.nz/tuia-encounters-250.

for one urban community is ill-being for another. Water practices are buried within the realm of embodied practices and domestic routines and can be understood only within a much broader 'messy terrain' of how water is made meaningful in relation to personal, social, civic, local and national identities (Allon and Soufolis 2006: 46).

How can an 'economic resource' such as water (largely framed by industry) be revalued in cultural terms, to help us understand what water in the city or in the countryside means to people and hence how the liquidity of mediated stories and memories of water can be redistributed? In Brazil, for example (as well as in other parts of the Global South), 'many cultural pursuits that have hitherto been described as "intangible cultural heritage" [UNESCO 2005], whether ceremonies, rituals, spiritual practices, are seen as continuous with an economy which includes a wide range of non-commercial cultural practices' (Bell and Oakley 2015: 157). Similarly, in the United States, Salvaggio et al. (2014) used value-belief-norm theory to establish relationships between people's knowledge about drought and water scarcity, general environmental attitudes and intentions to act. Place-based studies are critical because local context significantly influences attitudes into how people construct meanings about water, environment and their responses to scarcity.

Moreover, Somerville (2013) highlights how art forms and stories reveal intimate and local ecological knowledge of water embedded in contemporary cultural forms and languages in the context of the thirteen-year drought in Australia's Murray-Darling basin.[21] Somerville (2013: 85) even argues that the specificity of the indigenous knowledge for the region means that water is 'fundamental to Indigenous ontologies and epistemologies and the destruction of waterways and water sources is equivalent to cultural genocide'. Yet five years later the basin suffers its worst drought. Indigenous Australians in New South Wales have witnessed the Murray-Darling river system dry up completely during 2018–19 as temperatures rose to 46.6 degrees Celsius. Daryl Ferguson, a Yuwaalaraay educator, is quoted in the press as saying: 'Everyone says: "This is mine, this is my part of the river, this is mine, mine, mine." I'm not here to talk about ownership; that's not our culture. You want to own it? That's fine. Just look after it' (Davies and Allam, *The Guardian*, 26 January 2019). Therefore, research on cultural memories of water, watery senses of place, the specificities of water-based identity and psycho-geographies of drought are all well and good, but if findings do not connect with water science, water policymaking and water management, then these interventions will not have societal and practical impact. We need our science and policymaking to remember cultural memories of water.

---

[21] Photovoice and photography have been used by several researchers working with aboriginal communities, farmers and citizens on water use (see Maclean and Woodward 2012, or Sherren and Verstraten 2012).

5

Remembering and re-mediating women in drought

Narratives of drought as masculine

In *The Texas Water Journal* Ken Baake (2013) notes that policy and story become discursively interchangeable; with water users, the agricultural sector, property rights and policymakers as actors in an ongoing story in which fiction has an important part to play. Drawing on the 1973 novel *The Time It Never Rained* by Elmer Kelton based in the region and telling the story of protagonist Charlie Flagg (the archetypal individualist) in 1950s San Angelo, Baake argues not for the veracity, quality or meaningfulness of the novel but for its being 'important background reading for policy-makers' because its 'plain-written prose helps us [presumably Texans] understand deep-seated suspicion of government regulation in the name of the environment'. Thus, the book 'serve[s] as a literary exemplar of traditional West Texas values, along with the challenges those values bring to attempts at fostering environmental stewardship – particularly water conservation' (Baake 2013: 79). The complexity and paradoxes of these values pervade Baake's analysis, finding that 'the coexistence of anti-government attitudes with acceptance of subsidies at least among some producers exhibits a key finding in this research. All of us embody multiple perspectives [or mediated spheres] that at times are fragmented and paradoxical, modulated by expediency, pragmatism, and the need for economic well-being'. Moreover, he argues that '[k]nowledge of water and the aquifer is derived from multiple domains (science, history, religion, law, etc.)' again spheres and 'all these types of knowledge of natural phenomena and their impacts on people contain stories with plots'.

> Some of the most powerful of these stories are archetypal accounts of good and bad, cause and effect. [...] The federal or state government is easily portrayed as the enemy or monster at large. [Charlie Flagg's] suspicion of government agriculture programs and pity for those who took such aid was perhaps less borne out by an apocalyptic worldview and more out of the pragmatic belief that no one can better care for his or her resources than the person who owns them and depends upon them.
>
> (Baake 2013: 89)

It is in this neo-feudal and capitalist context of custodianship, private property rights and the *longue durée*[1] narratives of individualism, coveted land and freedom from governmental interference that environmentalism meets head on the stories and narratives that are most concerned with 'who' should manage water. The answer often seems to be, I would argue, *white men*.

Men in fluorescent waterproof protective gear, men behind complex statistical reporting on scenarios and men in engineering, chemistry, farming and business. This is more than simply a sciences versus arts social conditioning in favour of men more likely to be educated, trained and well-remunerated in industries that manage water on a daily basis. Controlling water has been culturally narrated as a masculine, imperial, colonial and mathematical pursuit but this does not mean stories of women managing water do not exist. In their research, Bryant and Garnham (2013) point to the gendered aspect of drought narratives relating to the pioneer myth. This had been explored by Stehlik et al. (2000) and Anderson (2008, 2010). Stehlik et al. (2000) interviewed men and women of farming families separately with the hypothesis that there would be gender issues. More than a 'masculinist paradigm of "the bush"'. Stehlik et al. (2000: 41) found that women also realized a role change, they became empowered and vital for maintaining *a way of life* that focused on the domestic sphere. Though '[s]acrifices were made for the "outside" [the farm] in preference to any "inside" jobs that some women felt needed to be done' (Stehlik et al. 2000: 47).

Deb Anderson[2] makes a similar discovery concerning Australian narratives of drought in which 'the outside' of rural Australia is imagined as a place to be conquered and controlled.

> Rural histories of the Mallee have presented spirited sagas of community perseverance in 'battling' a harsh climate [...]. Australian land use since the import of European agriculture has been a 200-year struggle to 'green' a brown land.
>
> (Anderson 2010: 68)

Therefore, what is striking about the remediation of *The Time It Never Rained* by Elmer Kelton, by both water stakeholders in Texas during 2011–13 and in Baake's paper that seeks to analyse the importance of the text in the collective consciousness of the region, is the gendered nature of the discourse. Drought in this context only seems to concern men, male protagonists, male rights to property, masculine pioneers and men's concerns with the state as emasculating.[3] *Where are the women in drought narratives?*

---

[1] 'Longue durée' is a term used in French historiography to define a longer-term narrative of barely perceptible changes that may be slow and work out over a long period of time. Drought has those features and as such is difficult to mediate as an 'event' in the way flood, storm and tempest are.

[2] Deb Anderson has published extensively on drought oral histories in the Mallee, bringing this work together in a book entitled *Endurance: Australian Stories of Drought* (2014).

[3] Likewise Australian narratives of drought are almost always situated in the over-arching national pioneer narrative, taming the country's wild, inhospitable semi-arid environments making them prosperous and plentiful (see Gross and Dumaresq 2014). This narrative is linked to individual and collective identities, such as 'Bush' character, steely, self-reliant and resilient, able to cope in difficult circumstances and environments (see Anderson 2014).

As a stimulus to this question, I mocked-up a newspaper article remembering the 1976 UK drought, created through free online newspaper software, for a group of UK drought researchers as part of the Drought Risk and You project (DRY 2014–18)[4] in order to make a provocative point for discussion that in a UK context, which does not have a *dry sense of place* compared to Texas or Australia, one should also assume that drought narratives are implicitly and explicitly gendered. In fact, I want to show in this chapter that mediations of drought by the press, television news, media archives and social media have framed drought stories through women and their bodies. All of which may block effective drought communication in the twenty-first century while making invisible real women's voices in drought histories, particularly concerning the Summer of 1976.

Remembering the 1976 drought

The UK 1976 summer produced a benchmark drought with days without rain ranging from forty-five to sixty-six days depending on location and is more likely to be popularly remembered as the greatest 'heatwave' in living memory if you were an urbanite; or, the worst and most traumatic experience for farming, if you lived in a rural location. There is a 'heat' to city living and urban culture as expressed by Thrift (2004) that is assumed not to apply to rural life:

> Cities may be seen as broiling maelstroms of affect. Particular affects such as anger, fear, happiness and joy are continually on the boil, rising here, subsiding there, and these affects continually manifest themselves in events which can take place either at a grand scale or simply as part of continuing everyday life.
>
> (57)

One such 'event' is a drought or a heatwave, and these are more than metaphorical, for one cannot assume that mediations of this event and its eventhood create and recreate a demarcation of the rural and urban as affectively distinct. As temperatures rise, maelstroms and feelings boil over, rise and subside, in locations that challenge the city's affective position as the de facto early warning system of environmental disaster.

Drought is the preferred term to reference for an extended period of hot weather in the UK's conventionally wet climate and is not 'on the radar' of the British public who assume that, put simply, as an island with much rain, the status quo of being wet and

---

[4] The Drought Risk & You (DRY) project began in 2014 and the author was one co-investigator. This £3.2m initiative was aimed at providing new evidence for managing future droughts based on science and experience with expertise in hydrology, geography, meteorology, agriculture, ecology, culture, media and communications from eight universities and research institutions. The project incorporated voices that would not usually be part of decision-making. Using a science-narrative integrative approach, which interweaves science communication with the collection of diverse drought narratives from a range of stakeholders, it aimed to democratize discourses and forms of knowledge related to drought.

damp is the default way of life.[5] This is a form of cultural amnesia and is not unique to the UK for it is startling that even in Australia during the record 2018–19 drought in the Murray-Darling river system of New South Wales, the Deputy Prime Minister Michael McCormack sought to erase a *dry sense of place* (and reinstate myths of flood) with a projected public forgetting by exclaiming to the press:

> The disaster was because it 'just hasn't rained. And when it rains, it will come down in such torrents people will probably be saying: "What are we going to do with all this water?" That's Australia.'[6]

Thus, it is the concept of the 'heatwave' (which displaces drought) that becomes the entry-point to a short-term, intermittent and unusual duration of hot weather for the 'British climate', the latter ordinarily narrated as summer weather to be escaped from by flying to locations where long spells of heat are guaranteed.[7] 'Heatwave' and its media representation provide opportunities for deeper and longer conversations and communication around drought histories and futures. But what exactly is being remembered about past British heatwaves and what new stories are remediating those memories? A recent 'hot spell' during the writing of this chapter in the summer of 2018 suggested an answer, with the BBC Breakfast News Twitter account curating oft-repeated archival excerpts from past BBC news items into a 'Remember 1976' news feed (Tweet posted 1:37 p.m. GMT 24 June 2018). Not curated as the year of the worst drought the UK experienced in the last fifty years but as #summer and #heatwave, moments for reminiscence and nostalgia.[8] The landing image for the one-minute video of archival excerpts began by remediating the iconicity of this drought event around images of young, lightly dressed women and hash-tagged the conversation in ways that invited reminiscence. While this may have had over 47,800 views at the time of writing (no doubt due to the opening image of women in bikinis), the many retweets and comments only served to collectively remember this period as one of outstanding weather, youthful love, pop music and socializing. The target audience of BBC Breakfast's Twitter feed being just the right age now to recall those teenage experiences of extreme hot weather; not within a trauma narrative but with an appreciation to the broadcaster for creating new opportunities to collectively engage in popular culture.

However, look beyond the current mainstream social mediation of 1976, and we can find that the popularity of the memories of 'heatwave' is countered by significant

---

[5] '"Watery sense of place" means [that] living with water and "water issues" (e.g. flooding) is part of individual and collective narratives of self and place' (McEwen et al. 2016: 15).

[6] Quoted in Davies and Allam (2019) 'When the River Runs Dry: The Australian Towns Facing Heatwave and Drought', *The Guardian*, 25 January 2019.

[7] See the DRY Project blog posting 24 February 2015 'Engaging the public on drought risks: stories about security' for the 'narratives in our heads' about climate and climate change. Available at: https://dryproject.wordpress.com/2015/02/24/engaging-the-public-on-drought-risks-stories-about-security/

[8] It is worth differentiating two types of nostalgia by drawing on Boym (2001) 'restorative nostalgia' of nation rebuilding and 'reflective nostalgia' of ironic, self-aware longing-ness. Boym sees 'restorative nostalgia' as placing emphasis on the *nostos*; she argues that it 'proposes to rebuild the lost home and patch up the memory gaps' whereas 'reflective nostalgia' 'dwells in *algia*, in longing and loss, the imperfect process of remembrance' (2001: 41).

and extensive shared memories of traumatic drought, particularly in the agricultural sector and rural communities, preceded by an unusually dry winter there followed a dry summer including sixteen consecutive days of temperatures over 30 degrees Celsius. This resulted in many reservoirs drying up, significant water rationing (even in a normally wet Wales) and a water deficit that continued for several months. Such an extended duration and significant severity of dry weather in the UK stands out for several reasons and not all of them pertain to hydrological history.[9]

While scientists in the DRY Project have undertaken much-needed modelling, data-crunching and scenario-ing, basing future probabilities of drought on past archival data from a rich and wide variety of sources, there is still a job to do to convince the UK population that the country is facing drought on a regular basis and that hot weather is more than spectacle and nostalgia. Particularly, if the UK drought discourse is pervaded by popular accounts of past hot passions, legendary heatwaves and urban youth cultures. As with the Texan cultural values that framed the beginning of this chapter, British cultural values permeate the drought narratives of the last forty years as an opportunity *to let one's hair down*. 'Think we're having a heatwave?', asks Brian Viner of the *Daily Mail* on 6 July 2015, 'Last week was Arctic compared to the sizzler of 1976: And those who lived through it will never forget it'. Proceeding to frame that lack of forgetting with the following sub-headlined bullet points which offer some serious *factoids*:

- In 1976 Britain was in the grip of a gruelling heatwave that lasted months
- At Wimbledon last week, temperatures were a degree hotter than 1976 record
- But in 1976, 400 people were treated in single day for 'sun exposure'
- Big Ben also suffered its first and still only major breakdown in its lifetime

This, unsurprisingly, is then followed by an image taken from the print media archive of a young woman in bikini top and cut-off jean shorts in a fountain.

With a weak repository of mediated and visual memories of the 1976 drought, wherein women's bodies stand in for serious remembering and women's stories and experiences, in particular, have been largely ignored in favour of sexualized imagery, it is no wonder that narratives of resilience have not been inherited or shared more widely. While we know, from the Earth Policy Institute (EPI) that 'heat waves are a silent killer, mostly affecting the elderly, the very young, or the chronically ill' (cited Bhattacharya 2003).[10] It is also these groups who are unlikely to be represented or appealed to in popular, commercial or public service drought media, even if we can find a way of unearthing and curating archival footage of

[9] It is worth noting that (at the time of writing) the hottest days on record in the Global North were in 2003 and recorded over 40 degrees Celsius in France. Thousands of people died across Europe as a direct result of the heat (with 14,802 of these in France alone). The UK record high temperature of 38.5 degrees Celsius in 2003 was exceeded in 2019 during the writing of this book.

[10] See 'European heatwave caused 35,000 deaths' by Shaoni Bhattacharya in which the author notes the EPI's claim that 'heat waves rarely are given adequate attention' even though 'they claim more lives each year than floods, tornadoes, and hurricanes combined' *New Scientist* 'Daily News' 19 October 2003 [online].

forest fires, water shortage and farming anxieties that have been uploaded to video sharing sites by local people, archival news pages and as anniversary material.[11] Therefore, paying close attention to drought storytelling outside of these nostalgic remediations of drought memories is vital if future drought communication in the UK context is to displace heatwave culture.

Drought storytelling

The UK DRY Project has explored how drought narratives might incorporate strong themes of blame, shame, deviance, punishment, control and anxiety, and it has also explored the overt and sometimes hidden and subtle politics in representations of UK drought and water scarcity from a wide range of stakeholders and communities: water companies, farmers, water users for leisure, heritage organizations, domestic consumers and environmental agencies. The project has revealed that narratives of drought, or rather, the daylighting of stories and anecdotes of drought alongside and in partnership with the scientific modelling, have allowed for an integration of science and narrative into an effective toolkit. This permits a thinking back through drought histories and forwards into drought scenarios in ways that not only afford water stakeholders the choice between hypothesis and observation but also allow for communicating water scarcity to a wider range of publics. The project has explored how drought is experienced and understood by different stakeholders (e.g. agriculture, business, community, health services, gardeners, water users) within the structures of UK Water Management, as well as in the everyday, embodied micro politics of water practices and a watery/dry sense of place. It has considered what responsibilities do different people feel they have when it comes to water saving, water management and drought resilience. Finally, it has explored the extent drought is understood as part of a broader environmental politics and a broader experience of water.

Yet, in tackling 'drought stories' in a UK context, my own research as part of the team discovered that everyday experiences and memories of drought were formed around the precise recollections of media and popular music during that hot summer of 1976, the discomforting feelings of pregnant mothers unable to keep cool, the shock of desertification of parks, grassland, favourite reservoirs and rivers, as well as the excitement of long evenings of late night drinking, socializing and how a drought offered a real sense of community and relaxation. The diversity of experiences of everyday stories of drought coalesce around more recent popular narratives that shape the past, and the Summer of 1976, as stuff of legend rather than the raw material for embedding environmental behaviours and inheriting new or improved relationships with water.

---

[11] See YouTube for '1976 New Forest fires' uploaded by user Chris Kirtley on 1 August 2008, which uses footage from Jack Hargreaves' *Out of Town* (Southern TV) to show the English country fires in the drought year of 1976. The BBC produced a webpage in 2012 with two drought news packages from 1976 of up to three minutes each: 'How 1976 drought left Britain bone dry' and 'Farmers' fears over drought' https://www.bbc.co.uk/news/av/uk-17100033/how-1976-drought-left-britain-bone-dry

Thus, one intervention into drought as a process that produces cultural, social and creative outputs, as much as scientific and governmental responses, is to consider the role of media in historically specific ways in a national and cultural context. More specifically, to understand drought's *memorability*. Such an intervention allows for a multi-modal and multi-perspectival approach that considers drought as a non-human 'actor' as well as acted-upon by humans, and it suggests we unpack drought representation and experience along certain axioms: urban-rural, public-private, young-old, men-women, weather-war and ability-disability as examples. These would not normally be within the purview of scientific approaches to drought history or to an iconic drought event, such as the period 1975–6 in the UK. In fact, they may be considered rather 'wishy-washy', a phrase I have heard more than once from the sciences towards arts and humanities approaches to what are global research challenges concerning water and extreme weather. Such a phrase denotes weakness and a watering down, but it also connotes the liberal arts as effete, feeble and indecisive. Thus, it might not be particularly surprising that the arts and humanities has tended not to address the problem of drought for fear of being not positivist and empirical enough, and furthermore that women's positions in drought histories and drought stories have been largely considered immaterial. Yet in 1970s Britain not only did women become especially visible and vocal, but it is in the summer of 1976 we can find evidence of one woman's voice as the main documenter of an everyday drought experience. That is the journalist, mother and farmer at the time, Evelyn Cox in her book *The Great Drought of 1976* (1978), which I will interweave in the analysis below.

In the remainder of this chapter I address key textual, visual and video recorded examples that capture the UK's hottest and longest summer of 1976, to date, that pertain to women's experience of water scarcity alongside the repeated representation of bikini-clad women as shorthand for extreme British warm weather as disruptive, festive and breaking with all the rules of convention and propriety. The current mediated memorability of UK drought is limited, in our living memory, to a handful of contemporary recordings of the experience through autobiography, video, radio, television news, newspapers and photographic stills. It is a slim multi-modal archive and my research at the BBC Written Archives at Caversham bore out my hypothesis that the media representation of drought has been rather limited to the domains of agriculture or 'blazing' news headlines.[12]

Undoubtedly, there is much more evidence in archives such as at the BBC's audio-visual collection (albeit not all of it is available to the public),[13] the Media Archive of Central England, ITN Archive, Southern TV Archive, Getty Images, newspaper archives and more. As a starting point to frame future more detailed

---

[12] While the BBC files pertaining to flooding in the UK were stacked on two trollies, the file on drought for the whole of the twentieth century was one slim collection of a few hundred papers. There is BBC (2012) *Drought 2012: An Inside Out Special* (TV Programme). In a recent visit to the Mass Observation Archive Sussex I noted little if anything had been collected on drought.

[13] At the time of writing Box of Broadcasts from the British Universities and Colleges Video Council (BUFVC) lists several free to air BBC programmes concerning drought, a good example being *Drought 2012: An Inside Out Special*, broadcast on 2 May 2012, which addresses the misrepresentation of London as having an abundance of water, considers learning from Spain's new consciousness of water scarcity through its education policy.

research of mediations of drought, heatwave and water scarcity in the UK, this chapter will focus on the most prominent example of the 1975–76 UK drought through and within which, I will argue, women emerge as both agents of resilience and remediated as dominant representations of passive sexuality. Thus, women's stereotyping wallpapers the collective memory of UK popular heatwave culture in unhelpful ways for water and drought communication in the twenty-first century. With typical iconic images of women in bikinis lying on scorched earth being taken from newspapers at the time or of female bloggers reminiscing about the hot summer of love, sex and vinyl records by posting images of themselves in bikinis, we have a drought memorability significantly working against the cultural changes necessary to represent UK drought today.

Considering the current scripting of the 'popular memory' of 1976 as a period of hot summers, passions and new styles of music, it is unsurprising, then, to see these tropes re-emerge in recent popular literature for the general reader, many of which come out on display in UK bookshops during a hot spell to signal summer or holiday reading choices. In Isabel Ashdown's 2013 book *Summer of '76* we find an 'intense novel of secrets and simmering passions' that 'takes us back to the legendary heatwave of 1976'. While, in Maggie O'Farrell's 2013 *Instructions for a Heatwave* we discover: 'A story of a dysfunctional but deeply loveable family reunited, set during the legendary summer of 1976'. In both cases, 1976 is marketed by publishers as 'legendary' and a 'heatwave' rather than a traumatic drought, while in the texts themselves the reader is provided with a sprinkling of science and policy to encode the meteorological context into the framing of intense domestic family life played out during broader social and cultural upheavals. Both books use archival data on temperature, water scarcity communication and drought to epigraph their chapters, with Ashdown's chapters beginning with a seemingly reliable Met Office report for that month. Each book, then, offers 1976 as a 'collective memory' on and through which to pinpoint, once more, a 'British' psyche tested and reshaped by external forces, just as in the representation of Winter Floods in 1947, East Coast Storm in 1953 and Summer Floods in 2007. On writing her book, Maggie O'Farrell states in the preface:

> Ask anyone who was alive in 1976 what they remember of that summer and it's as if a switch is flipped somewhere in their mind. The standpipes, they will cry, the swarms of ladybirds! The gardens scorched to shades of brown. The smell of tar from the melting roads, the forest fires, the corridors of A&E filled with severely sunburnt people. The cracks opened up in lawns. The heatwave plays an important role in Britain's *collective memory* of itself. There is a sense of unity in people's recall of it, a hint of pride in the way they all rose to the occasion and overcame adversity: such pluck hadn't been called for since the Second World War. Coming as it did in the middle of a decade defined by social unrest and economic instability, it represents a beacon of national solidarity in the face of an implacable enemy – the weather!
>
> (2013, my emphasis)

Here again, the Second World War 'keep calm and carry on' (a now global 'British' branding slogan) chimes with the 'Blitz Spirit' of interviewees in the *Sustainable Flood*

Memories Project (see Garde-Hansen et al. 2017) and remediates the war-weather nexus at work in many climate-related discourse.

That Ashdown's book misses the many months of dry weather from October 1975 preceding the Summer of 1976 is unsurprising, and in what follows I will explore how different media shape the temporality of drought in different ways. For example, the longue durée of a drought narrative is expressed in Cox's (1978) traumatic, chronological and detailed autobiographical account, which covers the devastation of the severe drought caused over a fifteen-month period on her rural English farm. In a different format, we have another longue durée of sorts but of an altogether sexist order of magnitude. This is a Pathé News' re-release on YouTube of film rushes of women in London in 1976. Erroneously titled *Heat Wave in London: Women's Summer Fashion (1976) | British Pathé* published 13 April 2014, it is clear that 'fashion' is not the focus at all, which I will analyse later. Over ten minutes and forty seconds of filming, without sound, the viewer's gaze (presumably a 'male gaze'[14]) is held within the frame of one long camera focus on female breasts and buttocks, a framing of the female body during a heatwave as passive material for drought narrative-building that does not push that narrative forwards, but places drought in a glorious and legendary heatwave past. While this nostalgic heatwave video creates the opportunity to frame women as scopophilic spectacle and their bodies as public property, it also invites a new currency of 'memories' and urban nostalgia to reproduce a narrative around the video amongst the YouTube posters. Here they reveal the 'imagined community' of 1976 and the 'collective memory' of O'Farrell's preface above, as one of longing for a lost past.

The third example concerns a different but equally iconic representation of older women or housewives (likely to be mothers) passively queuing at standpipes or carrying buckets to standpipes, in various TV news items, which are then remediated in newspapers and online. Such images predate 1976, as we can find them in the Pathé News archive published news footage from 1952 'Water on the Ration' in Sussex, wherein much of the footage shows women (dressed in housewife attire) walking to the town pump with their containers; and, from the 1959 'News in Flashes', containing film footage showing an older woman at the standpipe during a drought in Edinburgh.[15] In these early examples, film news footage (for cinema release) can be seen to be working hard to make a spectacle of drought, and they do so through these two iconic images that have become short-hand for drought and water shortage ever since: the dried and cracked reservoir bed and the woman at the standpipe, well or pump. Thus, we have footage of a dried and cracked Gladhouse Reservoir in the 1959 footage noted above, followed by a version of an externalized domestic scene, with the woman and the children, ordinarily 'in the kitchen', finding themselves out into the street at the standpipe.

---

[14] See Laura Mulvey (1999: 837) 'The determining male gaze projects its phantasy on to the female figure which is styled accordingly. In their traditional exhibitionist role women are simultaneously looked at and displayed, with their appearance coded for strong visual and erotic impact so that they can be said to connote *to-be-looked-at-ness*'.

[15] British Pathé (1959) 'News in Flashes'. Published on 13 April 2014. https://www.youtube.com/watch?v=MBFf5SD6Kek.

Likewise, framed as waiting happily, patiently and politely by news broadcasters, the 'keep calm and carry on' brigade of British women in 1976 recurs through the iconic image of women doing what many women around the world still do and will have done in the not too distant past: waiting and chatting at the public watering place. This reiterates O'Farrell's recollection of 'drought memory' above, women at the standpipes, while smoothing over the journalistic autobiography of Cox's rural account. In which, Cox stated: 'I soon realized that piped mains water is the most underrated convenience of the twentieth century. It must have contributed far more to the liberation of women than all the laws of Parliament have put on the statute books in the past few years' (1978: 30); and it is this wry aside that speaks volumes about re-reading cultural histories through environmental histories.

What we have, then, if we currently search online for media pertaining to the 'UK Drought 1976' is a visual representation of women that is either reminiscent of the misogyny around the 'heatwave' culture of the late 1970s or positions older women (often mothers) as inactive actors in water scarcity. In either case, women are (literally) thrown back in time to a time before liberation and emancipatory social movements of the 1970s. Consequently, women's unpaid domestic 'labour' and relationship-building that ensure a resilient community slide out of view, as emphasis is placed on their shrill and trivial chatter at standpipes or the spectacle of so many lightly clothed female bodies. While the imagined community of 1976 is deeply connected to the popular music of the mid-1970s, with nostalgia for 'girls' and hot summer nights, the reality in journalist-turned-farmer Cox's account presents the active and adaptable emergence of a powerful and entrepreneurial female subjectivity. It should not be forgotten that the Women's Movement was particularly active throughout the 1970s, when we consider the 1970 Equal Pay Act, the emergence of *Spare Rib* magazine in 1972 and the growth of Women's Studies across universities. However, the urban nostalgia for a youthful regenerative story of UK drought as one long hot and popular music-filled summer seemingly hides both the rural disaster of women's role in farming during this period and the older women as the maintainers of water supply in the growing suburban communities, such that joint pain due to carrying heavy buckets of water became recognizable female complaints at the time.

Therefore, in the rest of this chapter I seek to create a space for competing memories, images, videos and accounts of 1976 to recirculate. My aim is to demonstrate that multidirectional memory within a 'cultural memory studies' (see Olick et al. 2011) paradigm can orient people to past versions of drought from which they can derive ongoing and relevant personal and social resilience, so that certainty resides in a wider repository of fresh archive from which to draw new story material. This will make the current popular literature remediation of a 'collective memory' of the 1976 heatwave in which women's role is so one-dimensional, if revealed at all, more difficult to maintain. Moreover, it will admit into the discussion of social constructionist accounts of environmental resilience the notion of a lived body and lived experience negotiating drought, heatwave and water scarcity and, as such, influencing what is remembered and can be remembered about past droughts.

With the relative silence around the idea of drought in the UK (a nation built on an identity of wateriness), we should also add the invisibility of drought impacts, because images of drought and hot weather as iconic events are not as newsworthy as floods, which are dramatic and devastating, offering media organizations a 'spectacle' in a time-constrained and time-contained package. The longue durée of the story of drought is often imperceptible to fast-paced news gathering or requires story-ing over different, and often disconnected, media (literature, photography, video and online animations, for example) and through different media infrastructures (from analogue to digital, from local to national to global). While flood waters come and they go, disruptive events at a single point in time, drought is difficult to pinpoint the start and end of, and in our highly mediated environment is far less spectacular, with its slow and 'creeping' nature only to be imagined in terms of discourses of disability and invisibility.

It is interesting to note that when 'drought' itself is represented, the vulnerability it places on people and places is expressed in terms of disability. Take these examples from very different sources. The first is an online video from the academic professor Carolyn Roberts, the Frank Jackson Professor of the Environment at Gresham College, London, delivering a presentation on 14 June 2016, which was recorded and uploaded to YouTube. In it she defines drought in terms of its 'creeping paralysis' and the need to 'defeat it'. The second is from the now defunct print version of the British newspaper, *The Independent* (21 February 2012), which states that 'crippling drought' in the South East of England comes as a result of a dry winter and no forecast of Spring rains:

> Thames Water, which serves 14 million homes and businesses, confirmed there was a 'high chance' it would need to impose restrictions this summer as water companies struggle to replenish reservoirs, and rivers run dry or flow at a fraction of their normal rates, raising memories of the infamous 1976 drought.

In the article, a quotation from Mark Lloyd, chief executive of the Angling Trust, states that the 'vast majority of people are unaware that we are in the middle of a *crippling drought* – river levels are lower in many areas than they were in 1976' (Milmo 2012 [online]). It is not unusual to encode drought stories with notions of a quiet, slow, intermittent and barely perceptible movement towards increasing dryness, and a Google News search finds the terms 'creeping paralysis' or 'crippling' as frequently used to define a drought across journalism, science communication and academic writing.[16] This, of course, discursively encodes corporeal meanings into drought imaginings that positions the human-environment axiom as weak, frail, incapacitating, passive and paralyzed when exposed to water scarcity. Language is important here as any media or cultural theorist would argue, as do Mitchell and Snyder in their edited collection *The*

---

[16] For example, on the 16 June 2011 'Drought: A Creeping Disaster' *The New York Times*; while in another context on 3 May 2018 'Iraq Wilting: How Creeping Drought Could Cause the Next Crisis' Middle-East Institute http://www.mei.edu/content/article/iraq-wilting-how-creeping-drought-could-cause-next-crisis. In their chapter 'Drought: Pervasive Impacts of a Creeping Phenomenon' Wilhite and Vanyarkho make the point that it is the slow, creeping nature of drought that means it has been invisible to research funding agencies, governmental priories and news media: 'This is due largely to its slow-onset nature; cumulative, nonstructural impacts; low death toll directly attributable to drought; and extensive areal coverage. The large spatial coverage diffuses relief and recovery efforts' (2000: 245).

Body and Physical Difference: Discourses of Disability (1997). While on the one hand, to describe droughts in this way identifies a bodily, invisible and quiet deterioration that resists representation and aligns drought risk with those who are debilitated, it may, on the other hand, be deeply disabling in terms of public communication and response from those who seek to avoid narratives of insidious decline. If drought is described in terms of disability (crippling economic output as well as communities) and is frequently narrated as debilitating and 'in need of medical intervention and correction' then what is the role of science and government if not to intervene and correct? What is not considered in these representations, is that the drought experience itself (much like the disabled experience) 'is never imagined [as] to offer its own unique and valuable perspective' (Mitchell and Snyder 1997: 1–2), particularly one in which 'shame' plays an increasingly important emotional role in nudging individual and public behaviour (as noted in Chapter 3). With these ideas in mind, it is important to note, that drought storytelling is as much concerned with narratives of deviance, punishment, control and anxiety as it is with environmental concern and eco-consciousness. For example, in this 1976 news item (uploaded by Channel 4 News in 2012), the ITN archive footage on the water shortage and the use of standpipes addresses the first interviewee with the question: 'What's your reaction now you've been turned back on again?' The female interviewee (with received pronunciation) offers both an expected response of it being 'absolutely marvellous' and an unexpected response of irritation at the unfair use of standpipes in the neighbourhood, suggesting that the labour of gathering water should be experienced equally, paralleling this (again) with an analogy to shared feelings of resilience during the Second World War.

> We've all been rather upset to see that half the constituency are, sort of, on and we're off, and I think that, like the war, we don't mind if everybody else is the same as we are but you rather object to carrying buckets and buckets of water every day when you find that someone around the corner has got theirs turned on.
>
> (ITN Source: Drought 1976: archive pictures of the driest summer at
> https://www.youtube.com/watch?v=unJoZHD0AM8)

The labour of women's work during the 1976 drought (carrying water, caring for children and animals) has largely been subsumed under the popular media and cultural representation of crocheted bikinis and dancing in water fountains.

Spectacles of hot weather

Drought impacts to spectacles even if the slow and incremental onset of the drying of the land evades media representations. In the edited collection, *Extreme Weather and Global Media* (2015) by Leyda and Negra, they draw together examples which include heatwave and 'fire weather' as spectacular and here we find a very interesting chapter by Paula Gilligan entitled '"Blowtorch Britain": Labor, Heat and Neo-Victorian Values in Contemporary UK Media' on the cultural and social values articulated during the 1976 heatwave in the UK. While I would argue heatwave displaces drought in the public imaginary, Gilligan makes the point that heatwaves produce media-specific templates and offers the examples of news and advertising print media as opting for representations of women to stand in for expressing the extremity of British weather and connote the rarity of seeing so much of the female body in an everyday British context:

> The wildfire is the most media friendly of the extreme hot weather subset, and the one with which British audiences are most familiar through television. Print media are more invested in reporting heat wave stories, and appear to be fond of 'blazing headlines' on the subject: the high temperatures of a heat wave lend themselves to illustration with bikini-clad models and are particularly popular with the UK tabloids. Drought is not a specific subject that attracts much attention in the UK media outside of scientific circles.
>
> (Gilligan 2015: 100)

Thus, the 1976 drought has become most prominent in the collective memory of British weather narratives as more of a pleasurable 'scorcher' than a traumatic dustbowl with women playing a key role, even if their remediated visualization as semi-nude is placed alongside baked and cracked earth landscapes. When men do appear in these popular visualizations it is in terms of enjoyment of the blazing weather and blazing headlines, as for example, in the Guinness advertisement that played out that year, in which (as the camera pulls closer to the newspaper and the man's hand drinking the glass of Guinness) a soothing older male voice addresses the male audience with: 'This summer, sit back, relax, and enjoy the long, cold spell, cold Guinness. Phew!' (See 'UK TV Adverts during 1976' on YouTube). Therefore, curators of drought memory have an important part to play and a responsibility towards drought memorability in a changing climate if we are not to forget past drought impacts. While Getty Images provides many of the now-iconic photographs of 'warning signs', sunbathers and empty reservoirs, Pathé News, as noted above, provides the silent footage of women in London with the following 'Description' added by the archivist:

> Various shots of people walking around in T-shirts and shorts during the hot summer weather of the 1976 heat wave. The cameraman seems to focus on women. Lots of short shorts and tight tops. Everyone does hot quite hot.[17]

How people use archival media and media archives to remember 1976 is important to be aware of, and the access media producers have to media representations of one of the most severe droughts in UK history is equally important to critique. It's critical because we need to engage the public in drought through narrative methods

---

[17] Available on the official Pathé News website here http://www.britishpathe.com/video/heat-wave-1976, the video generates far more comments once it enters the YouTube platform. At the time of writing, the video had generated over 300,000 views and many comments which are memories – nostalgic, reminiscences, anecdotes, family stories, work stories and recollections of popular music and culture.

in which cultural change is at the heart of the process. We need to accept that the public are always already audiences of other media narratives as well as personal and family memories. The aspect gender plays in this is key to recognize not only in how heatwaves are 'sexed-up' as 'sizzling' but also how water scarcity impacts disproportionately on women as carers and particularly older women. In 2003, the heatwave across Europe had France dealing with a rapid increase in deaths, with women worst hit (see the BBC news item 25 September 2003 'French heat toll almost 15000)'. Yet, more recently, in the RT America Video 'California drought: rural areas hit hardest, wells infected with carcinogens' published on YouTube on 15 July 2016 the 186 comments mostly focus on how 'hot' the female news presenter Brigida Santos looks in the news report.

Conclusion: Towards dryness

Drought is, then, now widely recognized as a complex phenomenon due to the multiple aspects that concern its onset and, in the science community, the many 'types' of drought recognized; such as, meteorological, hydrological, agricultural and civic. While it might seem odd that media, communication and cultural studies should turn its attention to studying drought when much of the knowledge surrounding it pertains to the harder sciences of engineering, hydrology and ecology, it is clear that drought is not only a natural hazard. The implication and entanglement of class, race, gender, disability and social status with flooding as noted by Sayers et al. (2018) in which they identify those demographics more likely to be at 'flood disadvantage', also suggest that drought will be an increasing sociocultural hazard. Nor does drought only arise out of natural phenomena, there are anthropogenic factors, and the communication of drought is part of a wider and diverse communication of water issues. Thus, drought emerges (often slowly and incrementally) from a complex interplay between the climate (imagined as a system to be predicted, forecasted and modelled) and human activities (imagined as structured processes that can be influenced, systematized and nudged). Such a position, which may miss the complex messiness of unpredictable climates and the disruption of human emotions and social issues, is heralding a call for a better understanding of the 'agency' that human actors may or may not have in a context of 'Drought in the Anthropocene' (van Loon et al. 2016), particularly in the context of nations whose water supply is not publicly controlled. Drought research has then begun to appreciate the need for improved incorporation of wider societal impacts even if quantitative measures are expected (Bachmair et al. 2016). Moreover, there is also a recognition that while most studies focus on natural processes (e.g. Folland et al. 2015) or socioeconomic causes (Taylor et al. 2009), there is a place for the arts and humanities to explore the wider human relationship with water, as this book is undertaking, and more specifically the role of media, communication and culture in influencing preparedness and sharing knowledges (expert and lay). It may be unpopular to express this, but drought creates moral judgements (the 'haves' and the 'have nots', the 'wasters' and the 'savers'), which needs a critical Anthropocene approach as well as responsibilised humans.

In part, this chapter's focus on drought in terms of emotion, memory, nostalgia and trauma seeks to express new epistemologies of drought framing as place-based, gendered, affective and counter-hegemonic, and this chimes with James Workman's book *Heart of Dryness* (2009). As a writer and water analyst, he notes that 'We don't govern water' rather 'Water governs us'[18] and that revealing the powerful ways in which water is used in certain places to shape, influence and sometimes destroy people (as in the case of the Botswana Bushmen in Workman's text) only serves to erase the important ways in which resistance and resilience can be reshaped in a waterless landscape. The remaining Bushmen, whose water supply had been cut off by the state in 2002, survived 'one of the hottest and longest droughts in the region's history' in defiance, not only of the state but also of water researchers and writers, such as Workman himself, who had been focusing on modern, global, urbanized and technologized strategies (Workman 2009: 6). It is, then, the deeper, silent and quieter resilience of, often, ordinary people, and in the case of the research for this chapter, of women, that can be daylighted to reframe drought as socially and culturally constructed. As our perspectives on gender and sexuality shift, so too might our representations of a changing climate through recognizing hidden labour and resourcefulness.

In Mark Anderson's (2011) *Disaster Writing: The Cultural Politics of Catastrophe in Latin America* we are provided with an excellent arts and humanities approach to drought risk and resilience. The science of drought aside, what is key to his approach to Latin America is to reveal that water scarcity is not simply a matter of policy (scientific, economic, resource management or public engagement); it is also a matter of cultural policy, and by extension cultural narratives (which are nationally specific but globally shared, place-based and connective to other places facing the same challenges), and such narratives play a significant part in perception, reception and behaviour. Anderson's book has a very interesting chapter on 'Drought and the Literary Construction of Risk' and the introduction concerning 'normal nature to natural disaster' is also useful for those interested in exploring the tipping points as a matter of perception and hypothesis rather than only observation and modelling. His drought chapter pertains to Brazil and the late nineteenth and early part of the twentieth century, and he makes the case that as scientific data was absent, then literary data, folk culture and imaginative constructions served as evidence of drought and its impacts. Before science could make drought increasingly visible, art and culture served as the early warning system. Drought requires a persistence of vision that is difficult to represent through numbers alone, while stories are memorable, sustainable and inheritable. They may also be more inclusive along the lines of gender, race and class.

There is a through line in Anderson's book about an over-reliance on science and social science since the 1960s influenced by European thinkers permeating

---

[18] In a dispatch from Cape Town 19 April 2018, the journalist Eve Fairbanks also cites Workman's book in the context of water management in South Africa during one of the driest periods in which 'Day Zero' was almost reached (more on Day Zero later). She notes: 'Without some certainty around this critical resource – with its steady presence, largely hidden in industrialized society, made more unpredictable by climate change – society could fall apart', https://highline.huffingtonpost.com/articles/en/cape-town-drought/.

Brazilian academia, which places other kinds of knowledge about drought at a disadvantage. Anderson mentions several types of drought narratives at work in literary cultures that construct and ultimately institutionalize drought in the national imaginary (there was the Great Drought of 1877–9 and the drought of 1915 in Brazil as key events):

> These texts reformulated the vague notion of drought as a purely natural phenomenon of incalculable destructive force into a refined system governed by the interaction of classifiable variables, including social and political factors not formerly considered. More than impartial ethnographies, these novels' meticulous descriptions of local economic contexts, cultural customs, and political and social orders correspond to the calculated objectivity of risk assessment, with its aim of assigning contingent values to unknown quantities.
>
> (Anderson 2011: 66)

While working on drought representation in media and culture I am also struck by this 'assigning of contingent values to unknown quantities' that seems to be at the heart of both the science and art of drought risk and its representation. What is interesting in Anderson's analysis are the following points, which I summarize below, and which I have used in my media research with drought scientists in the DRY Project:

1. A literary formula adopts realist narrative tropes and collects the stories of 'typical' figures to stand in for the whole then adds facts so as to account for all of the causes and impacts of drought.
2. The tropes used by creatives focus on a character whose story stands in for family, community, nation, but the creatives were always outsiders to the experience of drought.
3. The stories show drought to 'be the tipping point', so it does not cause the social, economic problems but exacerbates pre-existing ones and the concept of 'moral drought' that precedes the actual drought plays a pivotal role in these narratives.
4. Drought narratives are not playful, satirical, humorous or use language in deconstructive ways: they are often dry and desiccated.
5. Drought risk is reduced to knowable probabilities (i.e. simplified) and the literary culture on drought serves the function of creating an essentialized view of drought in order for a calculation to be made.
6. Literature can set in stone drought discourses that continue to influence the national imaginary in specific ways either as a moral, racial, environmental problem or an economic, developmental, political problem, either drought is caused by the degradation of people and their knowledge or it is caused by technocrats mismanaging water. Perhaps both.
7. The relationship between drought and national culture and development are key, drought is seen to be either caused by too little or too much development.
8. Narratives of drought (past and present) serve to write places into a 'geography of drought'.

9. Policymakers will make reference to narrative tropes, past representations and stories of drought that can be so successful that they become institutionalized and thus drought becomes a political and technical problem rather than a cultural and social experience.[19]

10. Before policy, institutionalization and governance, there were locals who predicted drought, assessed its severity by carefully studying their environment through rituals, observing the behaviour of plants, birds, insects and animals, even 'numerology'. This is largely discredited and ignored by a science of drought. (Anderson 2011: 66)

Little of Anderson's ten points above permeate the communication strategies of water companies or water managers in the UK. Nor is it easy to find drought science taking on board point ten in its modelling. In the 'online backrooms' of UK water company communications departments some searching will reveal the more bureaucratic documents concerning water management and plans for drought. These cover modelling of drought triggers from the 1920s to the present day, and zone these triggers accordingly, with the severe zone defined as '*exceptionally low storage*', for example.[20] In such drought plans, water companies, for example, draw upon drought history to determine their trigger zones, noting that they expect to only restrict supply three times in every 100 years, based on past extended dry summer weather in 1944, 1976 and 1984, with a recognition that there were droughts in 1995–6 and 2011–12, and that such multi-year droughts are difficult to model, as is the future, which can only really be 'scenario-ed' (a term I will cover in more depth in the next chapter on flood risk and media).[21] In the UK context, 1976 is the key year here, and when asked by the UK Environment Agency to imagine future drought as more extended and hotter than this 1976 drought, the water company states that: 'A greater disadvantage of basing our drought planning on hypothetical rather than observed droughts is that we may then require huge investment for infrastructure that may never be needed. We consider this is unlikely to be supported by our customers' (Severn Trent Water Drought Plan 2014–19). Who then will be taking such bold moves based on such imagined scenarios?

We have the opportunity in drought communication to build in to our findings a recognition of the media and cultural narratives of drought and heatwaves that will compete for our attention with science communication. We should also be mindful of the collapsing of issues of racism, overpopulation or sexism into representations of a 'dry planet'. I have only touched on these in this chapter but in our reading of media texts we need to ask who or what is being blamed who or what is being blamed for

---

[19] See, for example, United Nations (2012) *Passport to Mainstreaming Gender in Water Programmes*.

[20] 'Storage is exceptionally low for the time of year. In this zone we consider, and potentially implement, drought orders to restrict non-essential demand' (Severn Trent Water 2014–19 Drought Plan).

[21] *Drought Plan 2014: Our plan for managing water supply and demand during drought*, Rightly, the Plan notes that no two droughts are the same, and the next one is not the same as the previous one: 'For example, there was a drought lasting from 1989 to 1992, three double season droughts (1933–34, 1975–76, 1995–96), a late summer, severe drought in 1959, and other single year droughts in 1921, 1984 and 2003' (Severn Trent Water 2014–19). But their modelling is based on previous ones.

drought and benefitting from heatwave and how does this distract audiences from the real victims and beneficiaries? Real stories of water scarcity in the UK have been silent and invisible; and even in countries with a drought risk, a watery sense of place is the ideological position to hold. If we continue within the current template of popular displacing narratives of heatwaves, where drought is for scientists and heatwaves are for the tabloids, we will indeed be taken by surprise every time. Water scarcity may well creep up, disable, cripple and paralyze communities, and many groups who are already disabled and silenced or invisible will continue to be disproportionately impacted.

6

Forgetting water: Developing a flood memory app

Introduction: Developing a 'watery sense of purpose'

Now that we have considered the remembering of drought in ways that remediate ideologies of gender, it is worth reflecting on what Leyshon and Bull (2011) defined as 'the bricolage of the here' or 'memory as it happens' as another form of social remembering. Here young people's stories of the countryside can be seen as the co-production of multiple experiences of spaces, places, identities, discourses, forms and practices. As Stuart Hall argued, the 'past continues to speak to us. [...] It is always constructed through memory, fantasy, narrative and myth' (Hall 1994: 226). It is in the re-building of 'the bricolage of the here' or 'memory as it happens' that this chapter operates and frames the stories, anecdotes and instances of mediating flooding, as opportunities to work against the forgetting of water. Media researchers have noted that digital technologies produce a surfeit of memory (see Garde-Hansen et al. 2009; Garde-Hansen 2011; Reading 2010; Mayer-Schönberger 2011; Hoskins 2018) as new ways of doing things remake old information. While scholars have been focused on what mediated memory in a digital age shows us as texts, cultures, sociologies, politics, identity, we also need to remember to 'read' media as 'media' and read them closely, if we are to understand the underlying economies of mediated memories.[1] These economies depend upon time, geography, industry, environmental exploitation and cultural value systems; and media create bubbles that protect humans from their stressful memories.

Memories of the UK Summer Floods 2007 demonstrate inter-generational exchange and learning and I consider how the design thinking process of developing a flood memory app (which is ongoing) allows for rethinking media for resilience and actionable knowledge. Workshops undertaken with the UK Environment Agency (EA) and Gloucestershire Heritage Hub between 2010 and 2019, as well as EA communication challenges posed to the researcher, and the use of media for flood preparedness alongside the challenges of 'forgetting flooding' (see Ullberg 2017) have been drawn into the development of the flood memory app. In order to develop an account of the co-production of flood memories with digital media further, I will

---

[1] Arts and artists bring skillsets to activate people, using the arts to break down hierarchies, speak to 'under-served audiences', which could address demographic histories that are changing. It is clear, for example, that flood-risk mitigation strategies are known, experienced and felt by communities but can arts, media and cultural production methods account for these?

consider, in this chapter, how a *trans-medial waterscape²* can emerge through memory apps rather than media archives (as covered in Chapter 2). As Paul Connerton has argued:

> To say that something has been stored – in an archive, in a computer – is tantamount to saying that, though it is in principle always retrievable, we can afford to forget it. And this forgetting becomes all the more necessary when the burden imposed on memory [...] becomes a problem for society in general.
>
> (2008: 65)

Sustaining flood memory is a significant challenge in cultural and social contexts that work against remembering, that is transient communities, business-as-usual, trauma and loss of living memory.

McEwen et al. (2014) report that creative and narrative approaches to flooding and water management – through community engagement, memory practices and historical narrative gathering – provide stories of water that can carry alternative knowledge reflecting a different kind of 'expertise' (McEwen 2007, 2011; McEwen and Jones 2012). So, then it makes sense to ask the question *who is socially mediating flood stories and with what purpose*? Narratives can also reveal different power dynamics at play within local water management (McEwen et al. 2014). We can observe that lower tiers of governance play a crucial role on the ground in terms of accessing and sharing local knowledge and feeding that back into larger scales. *What flood stories are being circulated and fed back*? Concerned, informed and active individual citizens are 'being supported to become part of the resilience and planning structures, by monitoring water courses and drainage systems at the most detailed, local level' (McEwen and Jones 2012: 685). How is digital media useful (outside of commercial interests and in between water events) *for connecting citizens into social movements on water management issues*?

A key theme that developed out of the above-cited research, of which I am a part across UK and Brazil water projects, is the idea of a 'watery sense of place', by examining narratives of how communities live with and celebrate water via historical and media archives, oral histories, photography and cultural and artistic resources. While these provide rich accounts of a distinctive landscape and lifestyles associated with a watery place and its community, how does one foster a 'watery sense of place' among those who feel displaced, out of place, transient or migrant? As Hood Washington et al. (2006) argue in *Echoes from the Poisoned Well: Global Memories of Environmental Justice*:

> Understanding and appreciating the knowledge embedded in the memories of communities that have been environmentally disenfranchised is critical to knowing more fully the social ecology of the world at large and the environmental costs of technological developments.
>
> (2006: xxii)

---

² This is characterized by a repertoire of discourses, forms, practices, artworks, creativities and imaginations.

On what basis, then, is a community being built if a 'watery sense of place' is only needed for problematizing dominant flood narratives[3] of risk and disaster that forgets micro-social opportunities? While engendering a sense of community and shared experience, and reconnecting communities with the rhythms of nature is all well and good, what of the global citizen who requires a participatory approach that is socially mediated but does not even know their direct neighbours? Can we detect what I will term a 'watery sense of purpose' among new kinds of digital hydro-citizens who use their personal memories and agentic capacity for a more mobile and agile sense of water, but need this connecting to micro-stories of flood risk and resilience?[4]

In *The Perception of the Environment: Essays on Livelihood, Dwelling and Skill* (2000), a book whose cover is a watery image, Tim Ingold makes a point that can be updated in a 'digital memories' environment (see Garde-Hansen et al. 2009). Here I am thinking about how new media technologies have emerged to socially mediate the natural environment in ways which challenge remembering institutions such as archives, museums and media organizations, but that leave the issue of *sense of place* (or sense of displacement) up in the air:

> If culture is taken to consist of a body of acquired information that is available for transmission independently of the contexts of its application in the world, then memory must be something like an inner cabinet of the mind, in which this information is stored and preserved from the vagaries of everyday life. Whatever people do, or wherever they go, they carry the contents of memory with them. It is an encyclopaedic resource which they can continually draw for guidance on how to proceed in a manner appropriate to the circumstances in which they find themselves.
>
> (Ingold 2000: 138)

While this *storehouse* construct of memory has been attractive to scholarship it does miss two key points this book is focused on. Firstly, the relationship between media and memory as 'mediated memories' (see van Dijck 2007) such that memory is collected, recollected, mediated and remediated at the intersections and interstices between agentic individuals and their social, cultural and natural worlds. Secondly, the co-development of media infrastructures and practices with the changes in the natural environment (as discussed in the previous chapter), wherein media do not develop outside of economic and natural resources (electricity, water, minerals, oil, plastics, seas, oceans and rivers for example) but inside and because of these resources.

---

[3] In earlier research, I contributed a conceptual examination of memory processes and the role they play in community knowledge(s). I highlighted how historical narratives of flooding can reveal a local and lived knowledge about flooding that is part of everyday life and maps on to contemporary experiences. This type of narrative is counter to the catastrophic disaster narratives that play out in the news media, finding that 'mediated narratives tend to create and re-create individual, community and institutional stories around flooding that are highly selective and determined by sub/urban-focused visual schema' (Krause et al. 2012: 130).

[4] See the previous chapter for more on the hydro-citizenship project and the special edition of *Ecology and Society* (2016) on 'Toward More Resilient Flood Risk Governance'. Available at: https://www.ecologyandsociety.org/issues/view.php/feature/115.

It becomes tougher to execute a global message of water stress through energy hungry technologies of smartphones, tablets, TVs and computers, all of which are contributing to emissions. Thus, a 'political economy' approach to the mediation of flooding, for example, would expect researchers to accept that (digital) media technologies, devices, practices and production cultures necessary to record, store, retrieve and recirculate content all have an environmental impact. This is not to reiterate a distinction between the natural and the technical but quite the opposite to focus on their inherent integration and co-constitutive histories (bubbles within bubbles). As Anna Reading noted (2014: 749), 'As yet, there is very little work within media memory studies that takes a political economy approach to the materiality of cultural memory and virtually no work as yet that seeks to conceptualize digital memory in terms of commodity chains of environmental impact, human labour, and material processes involved in the various aspects of production and consumption.'[5] In this chapter, I wish to work from the perspective of digital media and memory that does address the materiality of cultural memories of flooding and considers the labour, memory work, technology and processes, through undertaking a co-production model with flood-risk communities.[6]

Flood memory as metadata

What if, then, we conceptualized flood memory as a form of metadata? Traditionally flood memory in the form of metadata is 'structural' and 'material' as flood marks, flood gauges, flood levels on buildings, at bridges, churches and in the landscape. These material, monumental and official memories of flooding are both formal and informal, and they mark out the narratives, stories, anecdotes and affective nuggets of what we might see as data. Metadata is important for locating data, in the way it collects and points to other data and informally communities have marked flooding through plaques, photographs on display, through private albums, newspapers, and personal mementos/memorials, as well as through immaterial storytelling, anecdote and performative modes (see Figure 6.1).

New technologies for marking and remembering flooding have moved flood memory to online and digital spaces and places (artwork, photo-sharing, blogs, social networks and tweets)[7] that now describe and structure flood memory into new forms of metadata, such as the results from a simple search of 'UK flood' on Google Trends.

---

[5] Reading also makes the point that: 'The commercial rhetoric of the industries involved in the production of globital memory [i.e. digital and media industries] is the opposite of their commercial practice. Their rhetoric suggests that, unlike past industrial processes, these are clean, green, discrete industries with low environmental impact' (2014: 754).

[6] Ruppert et al. (2013: 28) make a similar point: 'Rather than occupying a "space of flows" or a virtual informationalized world, digital data is itself a materiality that is "alive," embodied and mobile' and we should avoid tending 'to direct attention away from the materiality and productivity of digital devices'.

[7] There is a body of research that explores the effectiveness of a topic-based *sentiment analysis* for the social web (see Thelwall and Buckley 2013). Such methods track reactions to events by collecting tweets, for example – and mapping positive or negative feelings over time.

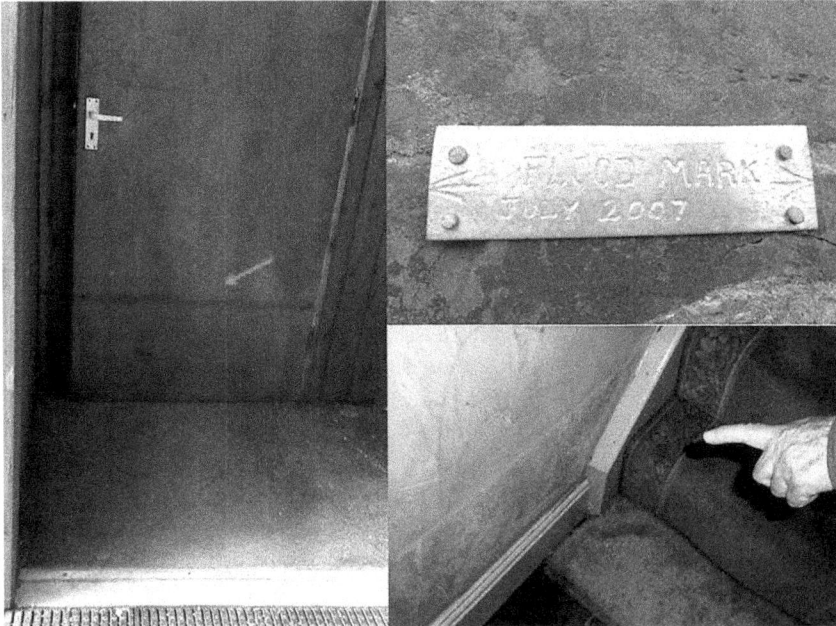

Figure 6.1 Unofficial flood marks of residents inside and outside their homes and gardens (*Source*: Sustainable Flood Memories team).

When we consider social mediations of water issues, the analysis of narratives and the complexity of water stories are enhanced by the fact that the communication on Twitter is happening in networked and distributed conversations. These are, often, single tweets serving as 'micro-content' or as 'nanostory' of 'transient bursts of attention' (Wasik 2009: 1) that are then wrapped in the perceived flow, stream or hail of information within a personal-networked timeline as well as in the spontaneous and ad hoc 'hashtag publics' (Bruns and Burgess 2011) that seem to collect and create reservoirs of stored stories. In the timeline, the stories are known only via social connections, but they become spreadable everywhere by sharing keywords and phrases. They are not simply 'fads', says Wasik, which 'conjures up too much of the unsavvy-media consumer of an earlier era' but rather appeal to a new kind of media audience who 'are so acutely aware of how media narratives themselves operate, and of how their own behaviour fits into these narratives, that their awareness feeds back almost immediately into their consumption itself' (2009: 4).

With this in mind, my research to develop a flood memory app with local communities that would store even the most mundane flood story, anecdote or nugget of remembrance focused upon key questions to address memory, risk and resilience:

- How do different communities in different national/local settings mark flooding over time? What is their relationship to flood marks (past, present and future)?
- How is flooding materialized and memorialized, are flood marks removed, is flooding forgotten/hidden?

- How can new kinds of flood-marking archives or repositories be built in these case study areas: as part of a toolkit for community resilience?
- How can memories be captured quickly in rapid-response catchments.[8]

If human response to flood has been found to depend strongly on prior flood experience, and I would argue to media's roles in flood memorability, then how quickly can those experiences be recirculated using new media methods, with voice-less communities at the centre? For it is worth noting, for example, that:

> Individuals with experience of flash flooding demonstrate a good understanding of its key characteristics. However, those who do not have experience of high velocity floodwater and debris do not necessarily demonstrate a spontaneous understanding of the danger flash flooding present.
>
> (Environment Agency 2009: v)[9]

Hence, the remediation of previous media recordings of flooding into new creative resources is increasingly possible as media researchers and creatives work more closely with flood-risk communities and scientists (urban and physical geographers, for example) using relatively new GIS-enabled platforms (such as ESRI, Open Street Maps) or through flood event games such as *Downpour!* (2016).[10] In the flood memory app development, we engaged communities in a friendly design of the interface (see Figure 6.2) adding opening script that would recognize the importance of memory for flood events.

Potential mediations of flood experience were presented as offering opportunities for submitting a digital flood memory to the repository and to anchor this in a number of ways: through geo-tagged locations, through references to extant media on YouTube or elsewhere, to upload a personal photograph or video and to story the experience as a memory. Moreover, if we take a media archaeological[11] approach to flood memory, as did the app development, then we do need to address not only the media discourses, forms and practices but also the media technologies of flood memorialization: using photographs, creating DVDs, remediating archival newspapers, video footage found, cut-up and re-made, digitizing museum artefacts, and using oral history interviews/

---

[8] In a UK context a 'rapid response catchment' is defined as a catchment that responds rapidly to a rainfall event. Key features of flash floods as defined by UK flood-risk management and response agencies are: flood severity, the short lead in time involved (usually defined as less than six hours); the short duration of the flooding; the link to heavy rainfall; dam failure as a possible cause; the volume and velocity of water involved; the danger presented by debris; the potential to cause material damage; and the urgent threat to life/presence of vulnerable.

[9] Thus, there is potential to extend and develop uses of social and digital media for collecting memories in rapid response catchments in the UK, in areas where there are longer histories of flash flooding to focus on community engagement and awareness raising in high risk.

[10] *Downpour!* was an interactive street game as part of Manchester Science Festival (20–30 October 2016) in which participants were asked to gather information, https://www.playfuel.co.uk/downpour, manage resources and protect the city. This is not too dissimilar to a BBC programme I often use in workshops entitled *Crisis Command: Flood*.

[11] See Jussi Parikka (2012) *What Is Media Archaeology?*

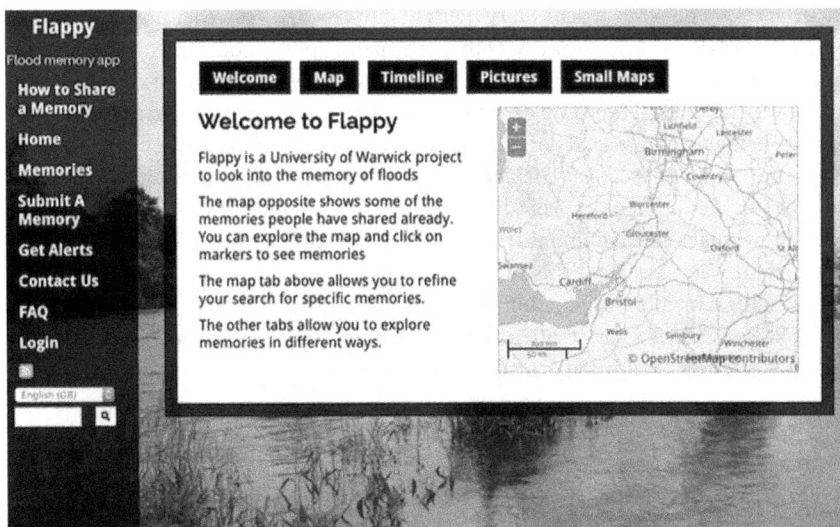

Figure 6.2 Screenshot to beta version of the flood app welcome page which uses Open Street Map and Ushahidi platform, and was the basis of workshops with flood risk communities for co-development.

digital stories as heritage for a flood memory app. Unlike typical broadcast media templates and representations of floods (see Chapters 1 and 2) that provide devastating, negative, out of control images with communities helpless and in need of rescue,[12] projecting flood memory as metadata can gather stories for knowledge, resilience and witnessing. These are dormant but accessible and shareable; and they work against strategic forgetting between flood events.

Digital flood memory work

A flood memory app allows users to remember the flood in terms of parameters (chronology, spaces, places and landscapes) and to draw on other media infrastructures, even the relationships between local and global are visible when we take the app to other flood-risk communities and engage new communities. Each community is involved in the tagging of their stories, determining what tags should be added and creating new ones for the moderators, thus providing valuable insight into values and the mixed ecology of personal, professional, structural and cultural scales at which they live. Communities that flood have the right to their flood memory (in all its mediation and memorialization), but they also have the right to privacy and to forget their flood

---

[12] Recent academic research that critically reflects upon floods and their representation addresses contested mediations of flooding within nations and regions during and after events. The Tsunami of 2004 (see Hastrup 2008); Hurricane Katrina in 2005 (see Littlefield and Quennette 2007; Robinson 2009); the Pakistan Floods of 2010 (see Murthy and Longwell 2013); the Brisbane Floods of 2011 (see Bohensky and Leitch); or the representation floods from the 1950s to 2000s in the UK (see Furedi 2007; Escobar and Demeritt 2014).

experiences,[13] and while it is *memory rights* that the media researcher implicitly evokes when undertaking *memory work* with participants, post-flood forgetting is equally important to acknowledge and work with. What is worthwhile remembering of the flood and at which points in time? What can we afford to remember and afford to forget? Will the technologies allow useful dormant flood memories to resurface when next needed?

In the *Sustainable Flood Memories Project* we took the following approach to co-creating flood memory (in dialogue), that is memories of flooding created with individuals and communities, towards the iterative production of personally meaningful water heritage. This follows a method of memory work that is a key part of the cultural memory production of the Brazilian digital Museu da Pessoa (Museum of the Person) which I have written about elsewhere (see Worcman and Garde-Hansen 2016), wherein the following principles are observed:

> *Principle 1. The idea that any individual can and should be considered a key actor in a global and a local (mediated) heritage – through the narration of his or her life story*
>
> *Principle 2. The right of every group to produce their own memory (emerging as a very strong policy – implicit and explicit – in Brazil and many developing countries in the twenty-first century, which parallels the focus on the people's memory in developed regions)*
>
> *Principle 3. The practice and outcomes of communities and individuals producing their memories, life stories and heritage, should be open, transparent and inclusive*
>
> *Principle 4. The potential for change – social, cultural and theoretical – should be a key outcome of (re)constructing and (re)performing the past in the present as personal memory*

Thus, flood events are not simply strategically forgotten[14] in order to move forward, but rather personal memories can be creatively produced through script writing, photo montage and digital editing into short films.[15] As Baake and Kaempf (2015) propose, narratives shape collective memories of recurrent events that occur within a community and past events can become guides to future events:

> Beliefs and attitudes that guide people's actions are reflected in shared local stories. When people have to decide whether they need to prepare for a major flood, they will stitch together what is known already about that flood event with stories of previous floods.
>
> (2015: 431)

---

[13] In my own research with flood communities I have been torn between the honour of listening to or reading a person's testimony of the flood event (knowing how much this would move audiences or provide knowledge) and then also honouring their request that I forget what they have said or written. It is too painful for them to have their flood memory shared, published (even anonymously), and produced in media for an audience. In an age of social media transparency and giving voice to those silenced this *right to forget* is a challenge for researchers.

[14] In our work with flood communities we found some deliberately forgetting floods because they felt that to remember floods de-values their property. Economics is playing a strong role here.

[15] See the work of Joe Lambert (2002) and the Story Center in California and the work of Meadows (2003) for more on the digital storytelling methodology.

Moreover, 'co-created memory' methods are one key aspect of how we approach digital media and memory work in ways that:

- recognize that media researchers, flood communities and water experts are active archaeologists of co-produced flood knowledge (i.e. memory-makers)
- are inventive, performative and offer non-deficit-model methods when dealing with humans and flood experience, memory and emotions
- ensure multiple-methods are employed for capturing the bubbles of mediated flood memory. For example, oral history interviewing for narrative memory; re-making home movies and family albums for visual memory; participant observation and co-research of Facebook pages for social and connected memory; digital storytelling for memory scaffolding; flood artefacts and memorabilia for sensory memory (sound, taste, smell and feel); or, making six-second memes with researchers for flashback and recurring memories.

Therefore, if researchers of flood history and memory accept the need for a community-led dig through media forms and practices, this will allow for a longer and more engaging view of flood remembrance and ongoing memorability. 'Memory long ago departed the cognitive limits of the body' argue Reading and Rossiter: 'Wax tablets, phonographs, X-rays, museums, newspapers, novels and cinema are among the many technologies that extend the contours of memory in highly material and culturally specific ways. Digital memory, especially mobile and social technologies provide yet new methods of capturing, sharing and assembling memories of the self and of societies' (2012: 5). Moreover, such an approach to digital flood memory work speaks to the 'digital [as] bound up with processes of re-territorialization, and the creation of new knowledge spaces, institutions, actors, devices and apparatuses' while at the same time knowing 'that these apparatuses draw from, or resonate with, older technologies' so rather than the constant emphasis on 'flows and mobilities, or epochal change' in media studies we should also attend to 'emerging stabilizations and fixities being performed in cascades of (partly social science) devices in particular locations' (Ruppert et al. 2013: 33) that are becoming increasingly connected and connective.

Connecting flood memories to water stakeholders

'What are the implications of connective and mobile memories for social transformations, for political engagements and public justice?', ask Reading and Rossiter (2012: 5) and one such social transformation surely ought to be our cultural attitudes to water and the need to transform our approach to flooding (both in terms of science and media). This is less a question about the cultural tropes and visual languages of flooding within 'state-operated institutions of memory [e.g. museums or the BBC] and the flow of mediated narratives within and across state borders' (De Cesari and Rigney 2014: 3) but more an articulation of flood memory as belonging to 'the person'. Flooding makes the person visible, if only to and with media. The growing relevance of increasingly digital and transnational *administrations of memory* finds new exchanges for 'communicative memory' (Assmann 1995, 2008) and new remembering policies and practices from networked actors in flood-risk management such as water companies, environmental agencies, local governments, NGOs, public service and community media, creative practitioners and heritage organizations. This can have the effect of empowering local communities and unlocking new levels of flood heritage through difficult memories. Working with environment agencies, NGOs and water stakeholders in the iterative design of the flood memory app, but not engaging those actors in endorsing it, allows it to operate in non-institutionalized ways. As one senior policy advisor noted:

> Such an approach demonstrates the 'art of the possible' in water issue communication. It provides a different perspective to what is currently being done and it can provide confidence for authorities to innovate in this area. It can be very hard for public bodies to trial and test new ways of engaging with the public as the risks of it going wrong are quite high hence they are often risk averse.
>
> (Principal Scientist and Research Theme Manager for Policy, Strategy and Investment Research, EA. Letter to author, 11 October 2018)

Therefore, I wish to make the case for a cultural heritage approach to digital flood communication, particularly in increasingly connected media spheres among increasingly disconnected communities. In Anheier and Raj Isar's *Heritage, Memory and Identity* (2011) collection, Ien Ang, the Australian cultural studies scholar, makes the point that diaspora unsettles 'the national'; it 'cracks open the nationalist narrative of seamless national unity, highlighting the fact that nations today inevitably harbour populations with multiple pasts, bringing memories and identities into circulation that often transcend or undercut the homogenising image of nationhood and national heritage' (2011: 82). This is particularly important to capture when connecting flood memories within and across flood-risk areas, for communities can homogenize around national and local heritage that excludes important actors and agents for change and resilience. This can lead to two key problems for flood-risk communication that developing flood memory applications can seek to address:

> *Problem 1: Young people are not engaged in flood memory.* Thus, projects are needed that evaluate the extent to which young people are engaged in flood memory as materialized through traditional practices (storytelling, flood marks, analogue media, flood heritage, for example) and new digital media practices (social media, online video, photo-sharing, tagging, geo-caching) and assess their intervention in this cultural memory of living with water, with the changing river and disruptive weather events. Making young people visible in mediating water.
> *Problem 2: Floods and Media can be disruptive to remembering.* Communities wish to forget and move on and media are not archived sufficiently or have a built in obsolescence.[16] Hence, identifying flood memorialization through social/digital

---

[16] We found in the Sustainable Flood Memories project that one local government interviewee referred to 'an elaborate archive as a Flood Memories or a thank-you book or something to commemorate the communities coming together to actually deal with this' but the book was 'probably lost'. Furthermore, digital archives can also be neglected and/or susceptible to deletion when handsets or laptops are upgraded, technology becomes obsolescent or there are changes to the media platform.

media applications would allow for new forms of flood heritage to emerge from these new communicators in the risk contexts and could have trans-cultural application, connecting those communities, and spreading memorability. It might also disrupt the emerging use of social media as only useful for crisis relief during the event itself and ensure a diffuse archiving of flood media.[17]

While there has been a consolidation of the contemporary 'disaster narrative' of flooding as 'a unique cultural genre of everyday life performance' (Carlin and Park-Fuller 2012: 21), there has also emerged an opportunity to produce transmedia narratives of floods. These could be in the form of flood games, books or apps, mini web series, animated movies and interactive theatre; or through exhibitions of flood memory on small mobile device apps with content (videos, audios, photos, interactive games); and, organized as water-related tags for multi-linear navigational flow, via connection among mobile apps and through museum exhibitions of physical objects. Retrieved flood memory research from local communities (as oral stories or mementos) can be used as interactive objects to navigate web versions of a flood memory exhibition, and this creates a communication protocol to enable the creation of several apps with different tools and technologies. Sensors placed on rivers, roads susceptible to inundation and in flood-risk areas can collect raw water data, and this data can be converted into tags and sent to the flood memory exhibition. All of this prompts and triggers more flood memory formation from audiences, users and readers enabling people to send their flood memories, which are tagged by the sender from the tags found at the flood memory site. This then provides the transmedia waterworld necessary for gamification such as a canalized river hunt game through urban streets looking for the source of the river and through this game participants learn how to take care of rivers, water sources and how to clean up their waterworlds (more on this in Chapter 8). Only now are we beginning the interdisciplinary thinking necessary to reconnect people to water, flood and drought into a mediated water universe.[18]

However, many people with these technologies available to them may be located in a nexus of social issues such as industrial decline and regeneration, low-income families, gender or racial politics. Deliberately low tech, digital/social media are becoming affordable and mobile in comparison with professional broadcasting standards. They can go places the professionals cannot and they are embedded in daily life. The convergence of increasingly user-friendly technologies such as scanners, digital cameras, video cameras, audio equipment, computers and mobile phones with the broadcasting capabilities of the internet means that the forums for hearing the voices of water stakeholders (not normally reached or included in flood-risk management) more possible. Thus, flood memory apps continue to develop in a new project in Brazil[19] using digital media and storytelling not to fix and freeze the past,

---

[17] As in the case of Deltares' and Floodtags' development of real-time flood maps using tweets in Jakarta 2015, http://fcerm.net/news/social-media-rescue-flood-events.

[18] See Groundwork Gallery's 2019 Programme entitled *Waters Rising: Making Art in Storm and Calm* 9 March–1 June 2019. The gallery sits directly on the River Purfleet.

[19] See the *Waterproofing Data* (2018–20) project in Chapter 9.

as media was once thought to do, but to make it mobile and travel across boundaries, connecting multiple audiences. Moreover, embedded in digital media technologies are different memory processes of pause, rewind, watch again, listen again, search, collate, thread, montage and archive, such that digital memory challenges us to think again about how flood history can be made and re-made. Clearly, the level of interactivity and skill that digital media encourages in the audience-producer means the passive audience who accepts the flood stories presented to it (as in Chapter 2) may become a thing of the past, and younger people's water stories may surface.

Conclusion: Strategic forgetting

Along with my colleagues, I have written about the concept of 'sustainable flood memory' (see McEwen et al. 2012a, 2012b; Garde-Hansen et al. 2016; McEwen et al. 2016). We offered an approach to flood memory as a form of cultural work. This community-focused, archival and mediated approach integrated personal and community experiences, involving inter-(vertical) and intra-generational (horizontal) communication. We determined that such memory was 'sustainable' in the sense that 'it creates and supports the conditions for its furtherance, acknowledges finitude and deletion, and has strong attention to intergenerational exchange and social learning. Thus, using associated lay knowledge in delivering on future resilience needs in relation to other forms of knowledge – particularly scientific and institutional' (Garde-Hansen et al. 2016: 60).

Connerton argues that there are 'seven types of forgetting' (2008),[20] and it is timely to consider flooding (disruptive and seasonal) as an environmental crisis (however locally contained) that is remembered and strategically forgotten simultaneously. The 'continuous process of forgetting is part of social normality' for 'much must be continually forgotten to make place for new information, new challenges, and new ideas to face the present and future', says Aleida Assmann (2008: 97). In order to move forward after the events, I have found in the research I undertook with the *Sustainable Flood Memories* team that recently flooded communities would on the one hand use media hyper-locally, producing flood messages and consuming them in a specific location for a geographically defined audience within a floodplain. On the other hand, other differentiated parts of those communities actively forgot flooding for personal and economic reasons, working against the need to mark the flood for future generations. The follow on flood memory app design and development sought to provide a softer communication method for materializing flood memories so as to find new ways of bringing a collective into existence, with increased possibilities for being social as a flood community (see Oliver-Smith and Hoffmann 1999; Jencson 2000). To speak of a 'flood community' is to address the realization that the community comes into being because of the flood and that the community, connected

---

[20] They are 'repressive erasure', 'prescriptive forgetting', 'forgetting that is constitutive in the formation of a new identity', 'structural amnesia', 'forgetting as annulment', 'forgetting as planned obsolescence' and 'forgetting as humiliated silence' (Connerton 2008).

by the flows, meandering, flux and force of rivers and water, can cut across socio-demographics, the 'urban' and 'rural', the regional and national, as well as expert and lay experiences. Flood memories and flood archives establish both a presence and absence of community, and importantly communities and their cultural heritage become defined by long histories of flooding in the light of the human design of wet landscapes into rivers, tributaries, pools, lakes, reservoirs and fenlands.[21]

Clearly, the materialization of flood remembrance (in photographs, flood marks, personal and official archives)[22] depends upon the social, cultural and material conditions and practices of the collective in the geographical area at flood-risk. It also depends more and more on media practices in everyday life, and on people training themselves in mediating flood memories for their cultural/social value:

> Actually, yes I did take some pictures. I haven't got them, they're on my old phone. Yes I took some pictures of it. I might have, maybe, uploaded them onto Facebook or something. I'm not too sure.
>
> (Male, 25 years, transient urban setting)

The flattening out of the flood-media-scape finds social media mapped onto water in ways that illuminate human interaction with the environment as mediatized but also acting as dormant repositories for flood memory. Within the dominant narratives of the mediascape, floods are recorded and represented nationally and globally as (almost biblical) human stories of natural disaster that may issue forth a politics of vulnerability, triumph or apocalypse. Equally media act as virtual flood marks and memorable data, such as tweets that report the location of flood water to become mappable along flood plains (see, for example, 'Digital trails of the UK floods – how well do tweets match observations?' *The Guardian* 2012). Within mainstream national/global communications articulation of the flooded environment, the human flood experience may fade from view if we only focus on the trace of that human story left on a data-scape. The combination of story and data, running in parallel, is necessary for addressing the forgetting of floods (between flood events) to ensure what is missing, hidden and forgotten (Muzaini 2015) remains accessible, retrievable and stored in ways that avoid media obsolescence. We also need to find creative ways of avoiding 'flood fatigue' where communities do not wish researchers to repeatedly dredge up their flooded past, thus respecting the tensions between flood remembering and flood forgetting for key individuals. Where memories and associated lay knowledges of floods are deliberately forgotten or buried, and there is evidence of the loss and

---

[21] Dilip da Cunha makes the case for seeing wet landscapes in terms of a very long human history of 'invention' from Alexander the Great to the Ganges in *The Invention of Rivers* (2019).

[22] There is a case to be made for more research to bring these media and cultural archives to the surface. Environmental agencies would benefit from the coordination and curation of historical media content to provide a narrative resource of past flood stories that draws on the archives. This could be used by media producers and water stakeholders and could be synthesized to create a timeline that brings together flood histories, generational histories, environmental histories and media histories. Many in flood-risk management have never mastered how to communicate the level of risk and likelihood of repeat flood to the public, and media archives can daylight both risk and resilience.

removal of material mnemonics that can strike at the archive's sustainability, then media have a vital role to play.

In 'Communities of memory and the problem of transmission' Pickering and Keightley (2013) define their appreciation of the work of Karl Mannheim (1959) as 'mnemonic imagination' that is 'a transmission of memory over time, in which the past is drawn into the present and reworked creatively in the interests of the future' (2013: 117). Thus, in the context of flood history, the vertical axis of remembering draws upon memories of flooding that pre-date climate change discourse and are important because learning to live with extreme weather conditions is not simply a contemporary phenomenon. Community flood memory has 'a vertical relation through time with what came before us and what may come after' (Pickering and Keightley 2013: 117). Hence, we can encounter historical and remembered changes in weather on a horizontal axis, being remembered in time (through modalities of archives, print media, oral stories, scrapbooks, anecdotes, home movies and regional news, many of which are creatively mobilized into new digital formations). When shared in the moment of, and directly after, a flood event (horizontally) or the event's anniversary, social media can reconnect these memories with the vertical axis of deeper time community memories of flooding. I would argue that media have the capacity to move memories, audiences and cultural values of water through environmental policy transfer across communities, as well as through the transmedial movement of emotional content between audiences.

Therefore, in the following final section of *Media and Water*, I wish to turn to the cultural policy and cultural values context in which old and new perceptions of water can be remediated to demonstrate how participatory and creative media and culture can be used to engage communities in re-imagining water futures. Remembering and forgetting flooding can have a significant impact on how flood-risk communities adapt and become resilient, there is little impact in exploring contemporary flood memory practices for flood-risk communities if the wider cultural policy is working against remembering by erasing the history of changing river levels, removing historic flood marks, ignoring lay knowledge of weather patterns and dismissing home-made gauges. Likewise, if media policies work against connecting, archiving and curating news, television, radio, cultural heritage, water festivals, amateur home video, photography and the social media remembering of water, then it is no wonder that audiences struggle to remember climate risks. Greening media is more than a matter of representation; it is about the matter and materiality of media themselves. In the final part of this book, I turn to the theme of perception and offer examples of sustaining water memories through digital data and digital story alongside remembering the urban river through inventive methods. New forms of water management, river regeneration projects around the world and the digital sensing of flood risk suggest that water stories are enjoying a renaissance of sorts. Yet these stories need to connect with data for water is also finding itself instrumentalized, channelled and cultivated through new and smarter forms of control; and media, communication and culture are playing multiple and sometimes conflicting roles in this process.

Part Three

Perception

7

The *cultural value* of water and water's impact on *cultural values*

Introduction: The cultural value of water

Bryant and George have recently stated in their introduction to *Water and Rural Communities: Local Politics, Meaning and Place* (2016: 1) that

> Water has been the subject of cultural thought and representation over centuries across diverse continents. It is used and represented in art, philosophy, religion and literature. Past and contemporary depictions and imaginaries of water, its surfaces and reflections, its depth and underworlds, have shaped ways of living with, and dreaming about, rivers, oceans, streams, rain, snow and ice.

They also argue that 'it is important to consider how the politics of belonging and activity' of water communities 'extend beyond national borders' (2016: 1).[1] It is, then, the mobility, convergence and globality of media and communications that offers some promising inroads for connecting water communities. Thus, applied cultural, media and communications research has emerged as a response to repeated attempts at engaging diverse, complex and sometimes resistant, peripheral or underserved communities in water policymaking (see Gibbons 1994 and Nowotny et al. 2001 in the science domain; and McGuigan 2004 and Giddens 1984 in the sociocultural domain). A 'narrative turn'[2] towards culture, media, creativity and communication in the sciences is not in conflict with excellence, evidence and data but recognizes that research and policy impact must engage non-academics, stakeholders, users, audiences and citizens. This is already bearing fruit in the social sciences, wherein narrative, media and story are playing a key part in understanding cultural, social and political identities, community-formation, risk perception, environmental learning and resilience strategies in water research and practices.[3] Yet culture, cultural policy

---

[1] They argue in favour of Askins (2009) concept of 'transrurality' to distance their argument from the 'transnational' which 'emphasises multi-scalar movements' and the complexity of water experiences (Bryant and George 2016: 8–9).

[2] See Christian Salmon (2010) *Storytelling: Bewitching the Modern Mind.*

[3] See, for example, Wiles et al. (2005) 'Narrative analysis as a strategy for understanding interview talk in geographic research'; Williamson (2012) 'Breeding them tough North of the border: resilience and heroism as rhetorical responses to the 2011 Queensland floods'; and, Tuohy and Stephens (2012) 'Older adults' narratives about a flood disaster: Resilience, coherence, and personal identity'.

and questions of cultural value are not necessarily the saviour the sciences can rely on for communicating environmental change through narrative, storytelling and lived experience. In this chapter, I draw upon theories of cultural policy and cultural value to address critically the narrative turn within water research towards perception change. This is as much about the protection of culture, creativity, expression and rights to remember and to forget, as it is about the protection of the environment.[4]

While there is 'a tension in that excellence tends to be measured by internal scientific criteria whereas usefulness involves other, extra-academic actors and partners' and that this 'is related to the boundary between academic cultural research and political, social and cultural practices found in other parts of society' (Forna et al. 2009: 8), we cannot assume that telling stories about water leads directly to positivist action. It is far more complex than that, as researchers of media and cultural studies know only too well; one cannot simply inject a water scarcity and conservation message, for example, into the audience and expect the intended meanings to be accepted. Areas of water policy (resources management, river/coastal regeneration, flood and drought risk management) which deliberately or blindly exclude the cultural from their purview because they are dominated by science discourse may find themselves out of touch with water users and those for whom water brings identity and being.[5] Yet, although this is where storytelling comes in, alongside media technologies for sharing and connecting stories of water, it should be with a critical eye on how (and whose) cultural values of water are being promoted, in which specific contexts of media and cultural production and through what kinds of activation of cultural and collective memories of water. This suggests an approach to water policymaking that is increasingly collaborative, participatory and creative, as Connick and Innes (2003: 178) noted some time ago:

> Many evaluations of collaborative policy making miss the mark because they come from the perspective of an older, modernist paradigm of policy making predicated on the assumption that policies can be designed to produce predictable outcomes, even in very complex settings. But collaborative policy making represents a different paradigm more suited to the postmodern, fragmented and rapidly changing information society.

Therefore, outside of the sciences and humanities research agendas (and their respective and almost exclusive focus on governmental and corporate agency to have impact) there are complex social and cultural lives whose intergenerational, emotional

---

[4] The UN *Convention on the Protection and Promotion of the Diversity of Cultural Expressions* (UNESCO 2005) seeks to safeguard rights to 'create, produce, disseminate, distribute and have access to their own cultural expressions' (Article 7) and 'take all measures to protect and preserve cultural expressions' (Article 8). 'Cultural expressions' are defined as resulting from the creativity of individuals, groups and societies and as having symbolic meaning, artistic dimension and cultural values.

[5] Bryant and George (2016) draw together key case studies of irrigation policy and river practice (in California and Australia), with cultural memory, identity and sense of place and community. Consider also the importance of rural values on gender, class and community or the gendered identities within a patrilineal system of farming, in order to understand how relationships to water engender marginal identities (see Ahmed 2005, 2008).

and inherited behaviours may resist policy changes and engage in water practices that demonstrate diverse cultural values. This helps explain how 'social learning', often evoked in water communication misses the mark if it only takes a social sciences perspective on 'the social', as Collins et al. (2007: 571) noted in their 'systems-thinking' approach to water management in the UK:

> While working on social learning we encountered different perceptions of what constitutes 'social'. Our policy interviews revealed a tendency for policy-makers to engage with discussion of the 'social' by developing a view that what is required is a 'sociologist', i.e. they identified an academic tradition as being responsible for 'people matters'. There was little evident feel for a more holistic or systemic view of the knowledge and skills that might be required to progress a situation that concerns a multiplicity of stakeholders and relatively complex communication processes.

Such complex communication processes involving media, creativity, culture, cultural values, digital technology, user experience, memory and emotions have been the focus of this book and here is where 'inventive methods' (e.g. participatory, action research and performative) are needed.

In recent years the social sciences and the arts have explored speculative research, topological thinking and complexity (see Lury and Wakeford 2012; and De Landa 2006), towards unpicking the hierarchies of knowledge in terms of 'assemblages' and 'processes'. This allows for a re-ordering and dis-ordering of knowledge so that the research innovation comes from the purposes to which the methods are put. With this in mind, it makes sense to challenge how knowledge is transmitted from elite professionals to a range of actors (through a deficit model) in tried and tested modes of address and to accelerate interaction, dialogue, selection, creative connections, actors and networks within, between and across cultures (academic, industrial, governmental, civic, social and personal). Approached in this way, participatory and collaborative methods are required, which maintain heterogeneous modes of address and do not erase cultural and personal knowledge. An inventive space for appreciating how different water stakeholders, users and communities change or resist change when under transformative pressures can be found in many creative and ethnographic methods based on narrative: (digital) storytelling, oral history, memory-work, community archives and heritage, family histories, anecdote, visual stories and social networking online (see, for example, Klaebe 2013).

This chapter, in making a case for changing perceptions of water and water management, introduces some key concepts: the *cultural value of water*, a *cultural policy of water* and understanding water's representation in terms of cultural policy studies that challenge the measuring of value itself.[6] It draws on literature from geography and

---

[6] In her polemical writing on systems thinking and quantitative measurements of value, the American environmentalist Donella (Dana) Meadows noted in *Thinking in Systems* that '[o]ur culture, obsessed with numbers, has given us the idea that what we can measure is more important than what we can't measure. [...] Pretending that something doesn't exist if it's hard to quantify leads to faulty models. [...] No one can define or measure justice, democracy, security, freedom, truth, or love. No one can define or measure any value' (2008: 176–7).

memory, cultural policy and cultural heritage in order to consider what kinds of water stories are oppressed, how are cultural memories and values of water expressed and made useful, how and where are children and young people engaged in remembering and experiencing water and how can we share narratives of water across generations through media and communications approaches.[7] It is largely underpinned by my own research in geography and memory studies (see Jones and Garde-Hansen 2012) and as a director of the Centre for Cultural and Media Policy Studies at the University of Warwick. McConkey says in the relationship to memory's role in the creation of narratives of self in place that 'memory searches for connections between present and past experiences in its desire to know, to evaluate, to make sense of life; and it searches as well for connections among the properties of phenomena in the objective world' (1996: 447–8). Moreover, as a cultural *place-maker* I would argue that water (as river, site of irrigation, city fountain or even as flood) acts out (or at least is engineered to act out) cultural values and becomes a key part of cultural policymaking, urban/suburban planning and rural community development. Water's availability or otherwise (as drought) 'forms a historic and continuing structuring narrative. Thus, here, memory is centralised as the process by which water, and by extension community, are given meaning over time' (Bryant and George 2016: 43–4).

It is the interdependency of water values and cultural values that motivate the final section of this *Media and Water* book, to address perceptions of water as they play out through *riparian media and culture* in the next chapter and *water scenarios* in the final chapter. As Ioris (2013: 323) has argued: '[w]ater valuation is, therefore, premised on a relational, holistic ontology, in the sense that values are necessarily interrelated and emerge out of concrete, politicized socio-natural interactions.' While this aligns with the key debates at the heart of cultural policy studies and its attention to cultural value and the valuation of culture, we must accept that this value is based upon pre- and non-human work (of water, oil and gas) which slides from view.[8]

Therefore, in his review essay 'The Torn Halves of Cultural Policy Research', Oliver Bennett draws attention to the two distinct and seemingly mutually exclusive pathways

---

[7] There are many examples of water research and policymaking beginning to intersect with cultural policy. Consider UNESCO's Memory of the World project and maritime/oceanic heritage. The PROTHEGO project (PROTection of European Cultural HEritage from GeO-hazards), which researches how weather and water as landslide impact on cultural heritage. The Outstanding Universal Value of Petra, Jordan (a world heritage site), is susceptible to flash flooding. Consider also the Venice in Peril NGO that engages in risk reporting, climate change and the impact of luxury cruise liners on water levels for cultural preservation. See also Harvey and Perry's (2015) collection *The Future of Heritage as Climates Change: Loss, Adaptation and Creativity*.

[8] In *Putting the Water into Waterways: Water Resources Strategy 2015–2020*, the UK's Canal & River Trust accepts that '[c]onsideration should ideally also be given to social and environmental costs and benefits' of the UK's canal and river system; but, in reviewing the methods for measuring this kind of value 'it was clear that any meaningful analysis would require extensive input from specialists in the fields of social studies and environmental economics. Added to this, there is still considerable uncertainty and debate around the suitability of these different assessment techniques' (Canal and River Trust 2015). That aside the Canal and River Trust has re-framed water in terms of 'natural capital' (re-purposing canal buildings, investing in wireless networks, and the formation of a Northern Powerhouse in the North of England).

of cultural policy research. Firstly, 'culture as a set of signifying practices and symbolic goods', cultural values or a 'whole way of life'[9] (Bennett 2004: 237) in which cultural policymaking is intended to create cultural citizens whose tastes and behaviours may be commercialized, managed and connected, as exemplified by Lewis and Miller's (2003) *Critical Cultural Policy Studies: A Reader*. Secondly, culture as something to be instrumentalized, administered, observed and policed by governments, ministries, arts councils, statisticians and research organizations as determined by Schuster (2002) in *Informing Cultural Policy: The Research and Information Infrastructure*. In Bennett's review of both sides, he argues that 'the field of cultural policy research is still defined by a shared commitment to investigating the conditions under which culture is produced, reproduced and experienced' and that researchers 'need to look beyond the confines of their own worlds' (Bennett 2014: 246). Such a world could be pre and non-human. Why is this important to water research, water communication and water policy and management? Because, water like culture finds itself caught between 'centre[s] of calculation' (Latour 1988) for measuring its value to society (which often fail to fully fix, observe and calculate) and its flowing through persons, families, communities and experiences as pre- and co-productive of culture. As Ioris (2013: 325) has argued

> On the one hand, mainstream valuations have been criticized for ignoring the politicized connections between environmental injustices and the variety of water values. Despite the fact that water studies tend to be an integrative field of study, the debate on water values has been either concentrated on the economic dimension of water management or centred on the macro politics and cultural aspects of water use. On the other hand, critical authors are often reluctant to deal with intersubjectivities and the diversity of feelings that underpin political action.

Again, torn halves, but this time within water research, and still on the issue of the measurement of value, valuation, values and evaluation. How, then, to approach these halves in studies of water and culture? In what follows I offer two spheres that subsequent chapters unpack in more detail with case studies. The first considers the value of media and participatory arts in water research and communication as a form of *riparian cultural value* bubbling up by *thinking and mediating water at the margins* and within the self. The second imagines technical innovation and approaches to *water scenarios* that stretch the capacity of flood data to encompass participatory storytelling (scenario-ing), feelings and compassionate methods.

Margins

Veronica Strang's oft-cited *The Meaning of Water* (2004) is an exemplar of the more holistic approach to the study of water and her more recent publications have continued to address cross-cultural meanings, sensory experiences and water's old and new materialities (see Strang 2005, 2009, 2014). Uncovering the encoding of meanings of

---

[9] Bennett is alluding to Raymond Williams (1989 [1958]) here.

water in, on and around the River Stour in England and the public experiences of those meanings at the river margins, Strang deconstructed 'the vision of infinite supplies' (2004: 197) in the impending global context of threatened resources, while noting the emotive identity positions taken up through water use of those she interviewed. With their well-kept gardens, washing machines, leisure activities, bottled water and metering as just some of the water practices around which people built a sense of self, family and community. 'The water flowing down the Stour is both natural and cultural' she stated in her conclusion 'responsive to a changing spatial, temporal, physical and ideational landscape. Its material qualities – its composition, its transmutability, reflectivity, fluidity and transparency – are inherent, but also responsive to context'. Meanwhile those in a relationship with this water have physical, emotional and imaginative interactions that 'render it mesmeric, sacred, comforting, stimulating, beautiful and fearful' (Strang 2004: 245).

All of this thinking about, in and with water is so marginal to the instrumentalist paradigm of measuring water's economic values that it offers potential new epistemologies for thinking through water's cultural values (values that are not easily quantified). Ones that also afford decolonizing and inventive methodologies and accept that the mediation of water may reveal interactions that are discomforting, ugly, polluted, diseased and dangerous. If NGOs and stakeholders wish and indeed need to daylight the hidden stories of water for science, policy and urban/rural regeneration to have the communicative impact they desire, then the *story-ing of water* by scholars needs to be taken *on board* (for cultural historians of water can also be described as *implicit cultural policy*makers).[10] As such, they too can strategically forget water at the margins or marginal water users.

Within feminist studies we encounter a re-thinking of cultures of water from the perspective of women's histories and narratives (from the margins); and women have long cultural histories[11] of lived experiences at the margins of water[12] in many cultures. Beyond symbols, Pamela Odih, in her focus on river restoration in the UK for her book *Watersheds in Marxist Ecofeminism* (2014), critiques the use-value of water from the perspective of patriarchal power relations, such that dominating water and women speaks to an exploitation of care and community. Drawing on Mellor's (2009: 255–6) assertion that capitalist-colonial enterprise has gained ascendancy through the marginalization of women, colonial peoples, labour and the natural world, Odih argues for 'ecofeminist waterways' (2014: 1). All of this chimes with Toni Morrison's argument in *What Moves at the Margin* (2008), wherein she intertwines art, the river,

---

[10] This accords with the work of Jeremy Ahearne (2009) on 'implicit cultural policy', that is, although the aim here is not cultural policymaking, the effect or impact would be because historians are ascribing cultural value through their selections.

[11] In their 2017 call for papers for an edited volume provisionally entitled *Women and Water in Global Literature,* Staniland and Jones posted a message online stating that 'the symbols and tropes of liquidity have long been connected to notions of the feminine'.

[12] Book and article titles expressing women in relation to margins and marginality abound and a simple online search supports this. In my research in Belo Horizonte, Brazil, community members would remind researchers of how women would gather at the banks of the Das Velhas (Old Lady's River) to wash clothes, sing songs and share stories. Much of this women's culture was lost with urbanization.

imagination and memory (with an implicit celebration of the female artist), such that there is a confluence of human and non-human memory:

> the act of imagination is bound up with memory.[13] You know, they straightened out the Mississippi River in places, to make room for houses and liveable acreage. Occasionally the river floods these places. 'Floods' is the word they use, but in fact it is not flooding; it is remembering. Remembering where it used to be. All water has a perfect memory and is forever trying to get back to where it was. Writers are like that: remembering where we were, what valley we ran through, what the banks were like, the light that was there and the route back to our original place. It is emotional memory – what the nerves and the skin remember as well as how it appeared. And a rush of imagination is our 'flooding.'
>
> (2008: 76–7)

Moreover, gender is not the only identity to be (re)surfaced in reconsiderations that decolonize water and cultural values of water in relation to marginalized peoples. In Astrida Neimanis's (2017: 1) *Bodies of Water: Posthuman Feminist Phenomenology*, she states that 'we live at the site [read margin] of exponential material meaning where embodiment meets water'. Furthermore, 'the various interconnected and anthropogenically exacerbated water crises that our planet currently faces – from drought and freshwater shortage to wild weather, floods, and chronic contamination' means that our bodies are engaged in a 'meaningful mattering'. Therefore, to think about cultural values through memories of water in terms of a marginalized social, cultural and political collective accords with the idea of 'regimes of memory' whereby 'history and memory is produced by historically specific and contestable systems of knowledge and power and that what history and memory produce as knowledge is also contingent upon the (contestable) systems of knowledge and power that produce them' (Radstone and Hodgkin 2005: 11).

There are, then, water knowledges being shared from the Global South that suggest an upturning of water culture research from the perspectives of Brazil (see e.g. the Manuelzao Project[14] on the Das Velhas River), New Zealand (the national seafaring commemoration of James Cook and the Maori for Tuia 250 Encounters[15]) and Australia (the accuracy of indigenous storytelling on sea-level rises thousands of years ago[16]). With them a post-colonial provocation to the story-ing of water by the Global North and its long histories of cultural and economic exploitation of water, the seas, oceans and rivers. Such *Epistemologies of the South* have been defined in the

---

[13] This is the thesis of Keightley and Pickering's (2012) *The Mnemonic Imagination: Remembering as Creative Practice*.

[14] Visit https://manuelzao.ufmg.br/at the Federal University of Minas Gerais and Its Worldwide Movement for Rivers.

[15] As it stated on the Tuia Encounters 2019 launch page: 'In 2019, New Zealand will mark 250 years since the first meetings between Māori and Europeans during James Cook and the Endeavour's 1769 voyage to Aotearoa New Zealand', https://mch.govt.nz/tuia-encounters-250.

[16] 'Revealed: how Indigenous Australian storytelling accurately records sea level rises 7,000 years ago', *The Guardian*, 16 September 2015, https://www.theguardian.com/australia-news/2015/sep/16/indigenous-australian-storytelling-records-sea-level-rises-over-millenia.

work of Boaventura de Sousa Santos (2014), who considers the impossibility of *saying the unsayable* from the perspective of most of the world (at the margins) who have been epistemologically excluded from defining the problems never mind finding the solutions to environmental and global social justice. 'Is the conception of nature as separate from society, so entrenched in Western thinking, tenable in the long run?', asks de Sousa Santos:

> The answer that Western thought gives to this question is weak because it only recognises the problems that can be discussed within the Cartesian epistemological and ontological model. [...] global capitalism has never been so avid for natural resources as today, to the extent that it is legitimate to speak of a new extractivist imperialism. Land, water, and minerals have never been so coveted, and the struggle for them has never had such disastrous social and environmental consequences.
>
> (de Sousa Santos 2014: 23)

There is, one could argue in water research, a certain luxury to using all these liquid metaphors cited throughout this book, a liquidity that is only possible in a mediatized environment of cultural privilege not cultural drought, in an imagined Global North context of infinite supplies, where you are more likely to feel freely and physically proximate to a well-cared for river than in the Global South. If one of the common phrases of a Brazilian community facing water scarcity is *a seca ou a cerca* (a Portuguese homonym phrase used often that means 'not the drought but the fence') then water is not only scarce in your thinking but is very obviously canalized, colonized, corporatized and controlled by structures well beyond your control. Thinking *without water* and thinking *at the margins* may be necessary to overcome the European privilege of only ever thinking *with* water abundance, to think the unthinkable for the times of scarcity and drought.

However, if water is 'cultural' and one can argue in favour of a cultural and media studies approach to water (as this book promotes), upon what terrestrial (or amphibious) marginal[17] grounds should we define the cultural value of water? Marx argued that 'the domination of the land as an alien power over men is already inherent in feudal landed property' such that land and nobility are fused (1844/2007: 61). Where was water in these materialist accounts of culture, the basis of which form much Western cultural and media studies research and teaching? Customs, currencies and cultivations are, in Raymond Williams's view of culture, all terrestrial (such as the buildings, mountains, landscapes, farms, cathedrals, fields, trees, walls and teashops that populate his 1958 essay 'Culture Is Ordinary'). Williams makes no mention of water as a signifier of ordinariness, and yet water is ordinary and should be available to the masses as much as the British television that Williams promoted at the time; a common culture, water should not be missing or mismanaged. Yet water as cultural in the many arts and humanities approaches cited throughout this book does suggest an approach to

---

[17] See Krause (2017), 'Towards an amphibious anthropology of delta life', and the examples of amphibious cities in Thailand and Columbia, which would probably be impossible to envisage in Europe and the United States even if places in the Netherlands or Venice, for example, may not survive the predicted sea level rise.

researching (at least the representation, symbolic meanings and cultural production of) water from a cultivated perspective, wherein water is valorized, spiritualized and rarefied or water acts as precious backdrop, a scape, context, facilitator or victim to man-made endeavours. Thus, water in popular culture for the general public has, more recently, promoted the idea of water as having literary, cultural and spiritual values and as these being under threat. See, for example, Steven Solomon's (2010) *Water: The Epic Struggle for Wealth, Power and Civilization*; *The Water Book* by Alok Jha (2015); or, Philip Ball's (2000) H_2O: *A Biography of Water*. Water may even be under threat from new water culture, if the rise of the plastic-bottled water industry marketed through well-being messages of well-hydrated celebrities is not more fully critiqued.

Such media messages could be said to place water in a position of high or aspirational culture for readers with social and cultural capital, an appreciation of water cultures as being a matter of social and cultural taste as with any other cultural expression. Herein, we might locate water's 'quality and excellence', its 'sweetness' and its 'light' which Matthew Arnold (2004 [1869]) attributed to culture, with life-giving properties that are foregrounded as fundamental to fully understanding, conserving, cleansing and respecting it in cultivated societies. Water as not simply an ordinary resource of ordinary culture, but water as both gentrified (spas) and cult past-time (wild swimming). Water is channelled in engineered pipes and drainage systems and through social and urbanized infrastructures, and has the capacity, if treated well to regenerate and possibly to uplift (if the community will comply) with the underserved neighbourhoods where high culture is rarely encountered, or the places where science communication of environmental risk and resilience fails to reach. The potentially immersive, dirty, leaky, destructive watery spaces of marginal, peripheral, edge-land and canalized waters are to be cleansed of their commonality and ordinariness if we apply cultural value to water.

This is a cultural economy approach to remembering water and imagining water pasts and futures in ways that speak to the cultural values of particular water users through new urban waterfronts not hidden cultural backwaters, from wetland reserves and waterside parks, to decanalized streams as well as through the creative re-use of water towers and the installation of multi-media water such as the Crown Fountain, Chicago.[18] The latter, by artist James Plensa, finds 'children of all types gather[ing] underneath waiting for the fountain to spit [from is mediated mouth]' (Clark and Silver 2013: 38); and, in Seoul, the Vision 2006 cultural and environmental policy of *Seoul Tomorrow* imagined a 'global city, a Northeast Asian business hub, a clean healthy and green city, a city with "culture as common and ever-present as air or water" (Tomorrow Vision 2011)' (Lee and Anderson 2013: 74). Here is a cultural policy of water that both remembers and forgets the pre-human work of water in a changing climate. How then to storyboard water futures and water's future? What kinds of storytellers do we need, include and participate with, and with what kinds of tools and techniques?[19] One starting point is to consider the concept of scenario-ing.

---

[18] See more on Crown Fountain, Chicago, at https://en.wikipedia.org/wiki/Crown_Fountain.
[19] See Jensen's (2015) evaluation of the use of storyboard techniques to draw coral reefs for conservation training in Mauritius.

Scenarios[20]

Water is not only culture and cultural but is a cultural agent acting upon the human, shaping the environment, and provoking new and ignored epistemologies of water values into being. Toni Morrison, as noted above, beautifully encapsulates the idea of the Mississippi River remembering its meandering, and the Australian researchers Allon and Sofoulis (2006) demonstrate in their research that understanding the common-or-garden 'everyday water' experiences will require more cultural studies approaches and methodologies that acknowledge water's agency. Hence, we need mutual respect across non-contiguous domains of water discourses, methods, epistemologies and research practices, running parallel like the banks of a river and stretching the concept of 'scenario analysis' to include meanings and values.[21] 'Scenario analysis', says Mahmoud et al. (2009: 798) 'is the process of evaluating possible future events through the consideration of alternative plausible, though not equally likely, states of the world (scenarios)' while the IPCC (2008) defines a scenario as 'a coherent, internally consistent and plausible description of a possible future state of the world. It is not a forecast; rather, each scenario is one alternative image of how the future can unfold'. Therefore, methods that are close, personal, intimate and yet representative and participatory must work in tandem with computational, quantitative, modelled and scenarioed water futures. Developing online platforms for media convergence to combine official and unofficial methods can support decision-making. Crowdsourcing citizen narratives to add to past stories and memories, promises an integrative storytelling method that might help to share knowledge, from multidimensional perspectives and afford 'multi-directional memory' (Rothberg 2009) of water cultures.

Researchers, NGOs, stakeholders and water uses are demanding new approaches to the study of water and oceans in particular considering their current circulation in popular culture facilitated by improvements in video-recording technology from drones as well as satellite imaging (see Steinberg's 2001 *The Social Construction of the Ocean*; Steinberg and Peters 2015 on 'wet ontologies'; and, DeLoughrey 2017 on new 'sea ontologies' and the 'oceanic turn'). Again, the Global South emerges as driving what Armitage et al. (2018) describe as a 'Blue Revolution', which I would argue is being connected through media, communication and transcultural exchanges to create resistance, cooperation and alternative water futures that seek to decolonize thinking about water (see also Dalla Costa and Chilese's 2014 *Our Mother Ocean: Enclosure, Commons, and the Global Fishermen's Movement* on Global South social movements).[22]

---

[20] 'Scenario', Latin for scene. In the Italian *Commedia dell'arte* (sixteenth–eighteenth centuries) it described the plot of the play or performance, sketched on a sheet and pinned to the back of the stage curtain so that the players could see the series of events, characters and plots that were to take place.

[21] Sharp et al. (2011: 512) report that on the *Water Cycle Management for New Developments* project, positivist (modelling) approaches and post-positivist (meanings/values) approaches worked best by running parallel in a complementary way, rather than integration across oppositional paradigms.

[22] Brazil's environmental policy has been key to this engine for change for some time, though the recent election of right wing Jair Bolsonaro with his well-publicized antagonism to the environment is concerning. Alternatively, the New Zealand Prime Minister Jacinda Ardern promotes Maori 'guardianship' of the environment. Hawaiian surfer and cultural theorist Ingersol shows how to read the seascape in *Waves of Knowing: A Seascape Epistemology* (2016) on the cultural values of water.

This cannot be only seen as *blue-washing* through corporate social responsibility or the *blue-ing* of research and innovation for neo-liberal academics, in the way that the *green-ing* agenda[23] was also seen to greenwash deep interests which maintained a certain status quo. Blue Studies' perspectives from the arts and humanities must, in my view, avoid scenario-ing the future of water as only of loss, trauma and painful cultural memory to seek out constructive connections with science and social science. The time is right for drawing into the modelling, observing, collecting and projecting of data on water, rivers, oceans, seas and extreme weather alternative contributions that analyse media, cultural and social movements to consider representations, labour, alliances, imagination, social justice, gender and race, for

> [t]he Anthropocene has catalyzed a new oceanic imaginary in which, due to the visibility of sea level rise,[24] the largest space on earth is suddenly not so external and alien to human experience. The increase in extreme weather events is correlated to the cinematic visuality of flooding and tsunamis, in which footage of a king tide in Tuvalu can come to stand in for the world's rising ocean. This new oceanic imaginary has inspired an increase in a body of literature, art, film, and scholarship concerned with our watery futures.
>
> (DeLoughrey 2017: 34)

In considering these futures, it should come as no surprise that social and cultural 'memory' and strategic forms of forgetting (see Paul Connerton's 'Seven Types of Forgetting' 2008) have emerged as perspectives from which to study water in the arts and humanities and have begun to enter cultural policy studies in terms of intangible cultural heritage and the desire to preserve past material and immaterial memories of water. This considers not only the eruption of water-related events (storm, Tsunami, hurricane, flood and drought) into (new) collective imaginaries but also the memories of water before the acceleration of urbanization. It also chimes with the way in which cultural and social memory is itself characterized as canalized, with hidden memories, knowledges and stories waiting to be surfaced.[25] To remember too much is to be flooded, and to strategically forget becomes necessary to identity formation or to moving forward, *bouncing back* or resilience (i.e. business as usual): 'Imagine the extremist possible example of a man who did not possess the power of forgetting at all', says Nietzsche, 'and who was thus condemned to see everywhere a state of becoming: such a man would no longer believe in his own being, would no longer believe in himself, would see everything *flowing* asunder in moving points and would

---

[23] See Maxwell and Miller's *Greening the Media* (2012) as well as research on greening organizational change, greening politics, the greening of the state and the greening of business. In *Oceanic Histories* (2018: 15) by Armitage et al. they note that the 'oceanic turn in environmental history indicates a larger cultural and political shift in which "blue" has, to some extent, succeeded "green".

[24] We do indeed need arts and culture to re-imagine sea-level rise as far more differential than the simplistic scientific 'bath-tub model', which imagines global rise equally distributed. Circulatory models of flows take hundreds of years, so we need to imagine scenarios of the shapes of water rise through media.

[25] It comes as no surprise that Robert Macfarlane's *Underland* (2019) on the subterranean landscape features accounts of rarely thought of water in terms of deeply hidden memory.

lose himself in this *stream* of becoming' ('On the Uses and Disadvantages of History for Life' 2011: 74, my emphases).

We have, then, within memory studies, a burgeoning area of research, new appraisals of 'hurricane memory' (see Mock 2018), 'sustainable flood memory' (Garde-Hansen et al. 2016) and drought histories (Waites 2018), all of which signal that like water research, the study of cultural memory itself is 'unbound' in 'its transcultural, transgenerational, transmedial, and transdisciplinary *drift*' (Bond et al. 2016: 2; my emphasis). The memory studies boom since the 1980s has been largely Anglo-European in its focus. Much of the debate regarding the study of cultural memory has focused upon the tensions between the individual and the collective, political/institutional histories and personal memories. Recent memory research has taken a socio-technological turn towards the operations and extensions of memory: writing, recording, storing, managing, sharing, retrieving and archiving. Researchers have emphasized the communicative media and modes of transmission that have extended remembering, produced 'prosthetic memories' (Landsberg 2004) or that depend upon mnemonic capacities (Keightley and Pickering 2012). The underlying cultural and political economies of Anglo-European perspectives can be contrasted with memory practices in Brazil, for example, where I have undertaken water research and considered the oral cultures, mixed-settled communities, deep time, indigeneity and democratized technologies that can produce affective politics and 'practical authority' in water governance (see Abers and Keck 2013).[26] As a means of advocacy and empowerment for marginalized groups, we need to expose the political and cultural economies of memories of water and its management. The extension of technologies of memory through (digital) media and communications (see Garde-Hansen et al. 2009) has fostered cooperative, inclusive and participatory cultures in many water management settings. Recently, research gathered by Assmann and Shortt (2012: 1–2) in their edited collection *Memory and Political Change* has explored memory as having 'great ethical and transformative power', and they highlight 'the impact of transnational movements' that 'shake communities and worldviews'.

In the UK, the AHRC has funded large research projects on cultural memory in the last decade (e.g. Memory, Place and Moving Image 2005; Managing Heritage, Building Peace 2008; Conflicts of Memory 2008–10), and all this has been framed by futures-thinking. In *The Future of Heritage as Climates Change: Loss, Adaptation and Creativity* (2015) David Harvey and Jim Perry argue that *the future is not the past* thus suggesting, as with the scenario-ing of water futures, that we must be creative, imaginative and accept that the traditional view that 'heritage conservation carries a treasured past into a well-understood future must be rejected'. For water heritage, and that includes the cultural memory of water, what 'we [choose to] carry forward is not simply the best of the past; in fact, it must be viewed as a dynamic expression of societal values' (Harvey and Perry 2015: 3). To carry forward myths and narratives of free and clean seas, oceans and rivers disavow the mnemonic capacities of water to carry the

---

[26] From at least the beginning of the 1990s, democratically organized memory archive projects in Brazil have been implemented in a wide variety of settings; in areas such as education, health, land rights and housing.

histories of human industrialization, habitation and colonialism that can be found within it. We must learn to live with the water we have created, the water that acts as the super worker to our creations, while acting on water's remembering of us, such as the River Thames's status as the most plastic polluted river in the world in 2020.

Therefore, *water scenario-ing*, beyond the sciences is more than reading probabilistic change factors for a particular region in terms of a particular time period or set of data. Change factors[27] such as rainfall or temperature historically read against past measurements and past events (such as floods and droughts) can be scenario-ed or imagined as future climates. We can ask hydrologists what will the rivers look like in the future from a *future flows*[28] perspective? Equally, we can ask river dwellers what kinds of stories of the river will be (or do they want to be) produced for those futures? Both kinds of scenario-ing have value, are valuable and can be evaluated. The second approach is a different kind of scenario-ing question that requires two creative capacities. Firstly, *visualisations through storyboarding* such as in creating a very simple online storyboard exercise with drought-risk communities in the UK can combine 'image and narrative' to 'contribute to learning because they act as mnemonic elements' says Negrete (2013: 202). Which, while simple in their execution and likely to create complex emotional responses, these can be insightful for risk communicators, as Roeser notes, 'in risk communication, emotions should not be abused for manipulative purposes; rather, they should be seriously addressed in order to trigger reflection. [...] emotions can enable moral reflection and deliberation' (2012: 1037).

Secondly, the capacity *to visualise water uncertainties* such that several (and perhaps very many) equally possible futures (that do not necessarily agree) are created, transmitted, shared and debated as stories, will be key to integrating science and narrative approaches to water communication; and to developing hydro-citizenship for the future. As Wynne argues in 'Strange Weather, Again Climate Science as Political Art':

> to act as if politics cannot change the moral and behavioural outlook of citizens, by identifying their more generous and relational human spirit and giving collective articulation, hope and public presence – and material policy form – to these elements of human nature, is timid, wrong, and maybe even self-defeating in this domain. Elsewhere. I have suggested that mainstream social science tends to reinforce an atomized and instrumental, rational choice self-interest model of the human subject, and while this is real for sure, it is not at all exhaustive.
>
> (Wynne 2010: 299)

In the chapters that follow I offer two case study-based approaches to mediating water and using media to produce new perceptions of water through participatory

---

[27] There are a multitude of diverse change factors: land use, demographic, migration, governance, role of NGOs, social movements, political change, cost, efficiency measures to name but a few.
[28] See the Centre for Ecology and Hydrology's *Future Flows* project, https://www.ceh.ac.uk/services/future-flows-maps-and-datasets.

co-production models of river stories and flood-risk data. Both of which seek to challenge the instrumental, rational choice self-interest model of the human subject, offering instead a messier articulation of a watery sense of place. In each case, the cultural values of water play out for young people in two contrasting contexts, the UK and Brazil, where different cultural policy initiatives and different perspectives on citizenship afford alternative perspectives on how to mediate water.

8

Riparian media for marginal communities

A river's right to be remembered

Rivers dominated the global news during the research for and writing of this book. On 5 November 2015 Brazil's worst environmental disaster event saw toxic mining waste polluting the Rio Doce, drawing global media attention to river-human relations.[1] National narratives of climate and environment may be distinct in Brazil and Dayrell and Urry (2015) conclude that Brazil's climate change prioritization (at least during the last government) and the perception there is a general lack of climate scepticism challenges a 'modernization thesis', for the hope is post-industrial societal-cultural values will lead on resilience and community adaptation. At least, these are the expert-led research findings of academics; but, again in Minas Gerais a dam collapsed in the town of Brumadhino on the 27 January 2019 with hundreds missing, suggesting a disconnect between policy, research, commerce and practice on the ground. Concepts of sovereignty resting with people and participatory engagement (Avritzer 2009) may well hold promise for generating new forms of interactivity, governance and power-sharing in 'greening Brazil' (Hochstetler and Keck 2007), but are the public really included in river management decision-making here or anywhere else in the world for that matter?

Thus, media reporting continues to hold power to account as well as create myths on the mismanagement of water in many countries, while simultaneously being a tool of communication for vested interests in water exploitation. This suggests that if there is a framework of hydro-citizenship at work, it needs to be more digitally connected and globally mediated. It must also be inclusive, particularly of indigenous peoples who may not have property rights to a river but do have cultural rights and see their cultural memory being destroyed. It is clear, at least from my own research experience of working with communities who wish to engage medially, culturally and socially with rivers, that the 'intensely scientific primary framing' of Western environmental issues and the 'entirely economic imagination of appropriate responses' alienates ordinary people, at least in the UK (see Wynne 2010).[2] While urban rivers may offer

---

[1] There are many photos of the Krenak tribe next to the Doce river, one month after the Fundão dam disaster in 2015, circulating online and in news releases.

[2] For example, during a Walk with Water tour delivered by Talking Birds and a workshop delivered by the Warwick and Coventry universities (30 March 2019) alongside the River Sherbourne, Coventry, we uncovered stories of love and passion for the river amongst fairly marginalized and underserved community members.

corridors for sustainable living (see the UK's *URSULA* project 2008–11), when they are hidden, neglected and/or engineered, then new modalities for discovering, daylighting and sharing urban riparian stories across and between cultural niches are required. Put simply, *rivers demand to be both remembered and forgotten.*

On the other side of the Global South some kind demand was being met. On 17 March 2017 while researching the connections between memories of water and how communities use cultural memory and media to represent their deep connections with rivers in the UK and Brazil, I received this email (see anonymized extract below), that sought to establish the non-human rights to memory of a river in New Zealand:

> *Tēnā koutou e hoa ma,*
> *In case you hadn't heard I thought I'd share this with you all – ignore all the racist comments!*
> *http://www.nzherald.co.nz/nz/news/article.cfm?c_id=1&objectid=11818858*
> *See also*
> *http://www.nzherald.co.nz/nz/news/article.cfm?c_id=1&objectid=11819442*
> *For Māori this is a significant event, our Government has finally recognised a river as a person, and specifically they have given the Whanganui River the legal status of a person, it now has all the rights, duties and liabilities that come with personhood.*
> *Nga mihi*

This long-running Maori-led campaign had revealed two significant challenges to media, culture and communication research of the environment, cultural memory studies and communicating water science. The first is concerned with epistemology. As touched on in the previous chapter, epistemologies from the Global South can be seen to be challenging orthodox approaches to water management,[3] to cultural memory and to solidarities beyond national containers. Not simply in the case of participatory hydro-citizenship in Brazil but in questioning the reinforcement of the global capitalist status quo and the deliberate forgetting of rivers, streams and brooks in the pursuit of *business as usual.* It questions both the rationalist consumer approach and the water as natural resource approach, even if it does confirm the growth in directives towards recommending participatory and story-based techniques such as in the European Water Framework Directive (see Pahl-Wostl 2008). Media and creativity can, then, play an important active role in challenging perceptions of both the commercial interests of water companies and the indigenous interests in water cultures,[4] but they also act as conduits and platforms for nationalism, racism, sexism and xenophobia to be surfaced in the cultural conflicts around water.

[3] For more about the river and its changing status, it is useful to see updates on Wikipedia at https://en.wikipedia.org/wiki/Whanganui_River. Such indigenous rituals are not confined to the Global South, rather they are spreadable. For example, see this film that research collaborators in Bristol and Bath organized of the performance the 'Oath' to the Rivers Avon and Severn in Bristol on 22 March 2015. Available at: https://www.youtube.com/watch?v=u79gnQD39x8&t=178s. Accessed 2 November 2016. See also the *River Avon* project, Southville, Bristol, where participants draw maps of resilience, community agency, vernacular knowledge and water stewardship; and, there is *The Big Blue Map of Bristol* project using text, drawn images, journaling and photographs.

[4] The Declaration on the Rights of Indigenous Peoples (2007) seeks to secure the right to 'maintain and strengthen their distinct political, legal, economic, social and cultural institutions' (Article 5).

The second is concerned with the challenges to the 'national' container for both human memory and human rights to be remembered and the rights of the non-human actor (the river) to remember and be remembered by humans. Again this is facilitated by the digital connectivity of media for networking disparate research and cultural projects through what Reading and Notley term the materiality of globital memory (2015). In the context of remembering the river this would be to use digitally mediated memory to acknowledge cultural memory as materialized by water, dams, river banks, streams and to 'resist metaphors, narratives and concepts that attempt to remove digital memory from its material consequences' so as to 'incorporate an understanding of memory's materialism' (Reading and Notley 2015: 511). Alongside Paul Thompson's *The Voice of the Past* (fourth edition 2017), Joe Lambert's (2002) digital storytelling movement from California to the BBC,[5] the digital Museu da Pessoa (Museum of the Person) in Brazil,[6] the call for a right to memory in Canada,[7] and the proposal for a museum of the everyday from the Turkish writer Orhan Pamuk,[8] we have then movements spread over the world seeking to connect remembering to materiality. These offer everyday rights to memory that focus on cultural and media flux, mobility and identity, underpinned and facilitated by digital media platforms. 'Memories are also trans-national and trans-cultural', says Reading, for it is not 'possible to contain memories within national boundaries. Memories are mobile: they travel with people and without people; different memories rebound off each other' (2010: 13). Yet as Wynne (2010: 299) has argued for 'many people, apocalypse has indeed already arrived, and conditions which have been imposed on them by past and present – often environmentally damaging – global economic and political arrangements force them as a matter of sheer survival to do things which may well exacerbate climate and other environmental processes'. It is this apocalyptic narrative (one that cancels a pleasant dry future and, with it, hope) that is being challenged by the micro-narratives of well-being and participatory hydro-citizenship, which I explore below.[9] While fiction such as Iain Sinclair's *Downriver* (2004) imagines a River Thames of traumatic memories as an urban wound, we also have new fanfic-led narratives such as Ben Aaranovitch's *Rivers of London* (2011) series, with the Thames catchment as creative, gendered feminine, racially inclusive, funny and hopeful; thus, offering new ways for young people to understand a watery and amphibious sense of place.[10]

---

[5] See the Story Center, California at https://www.storycenter.org/.

[6] See the Museu da Pessoa, Brazil at http://www.museudapessoa.net/.

[7] See all the articles in *Media and Development* journal LVII: 2 (2010).

[8] Pamuk states that the 'aim of present and future museums must not be to represent the state, but to re-create the world of single human beings [...]. The future of museums is inside our own homes' (Pamuk 2012: 56–7).

[9] Science fiction and cult film have a long history of presenting apocalyptic and terrifying visions of humans and human habitations consumed by liquid forms from *The Blob* (1958) to *Princess Mononoke* (1997) to *Under the Skin* (2013) as key examples.

[10] Cultural and media engagement with the Thames extends in many different directions. For example, *Tales from the River Thames* (2012) Unicorn Theatre, London. The online oral history project of the lightermen and watermen in *Tales from the Thames* at http://www.thameslightermen.org.uk/. There is the *Totally Thames Festival*, which launched in 2014.

Therefore, this chapter draws upon new research undertaken on and with Groundwork, a UK NGO of franchised environmental charities, that seeks to work with communities to revitalize green spaces and waterways through corporate social responsibility, volunteer work with young people, river clean-ups and water monitoring, all communicated via social media and video. The use of media in river wade walking to make visible the city or riverbanks from the river's perspective gives the river a tone of voice, a point of view, a soundscape and a visual scape to represent the human back to the human. In the context of the analysis of Groundwork's *Jordan* video, which will follow in the later part of this chapter, there should be the wider appreciation of the national and transnational cultural programme of river daylighting work such as Thames21's use of media at the River Ravensbourne, London[11]; the collecting of stories for *The Manuelzao Project*, Belo Horizonte, Brazil; and, other river art/media projects in São Paulo.[12] Visually well-documented NGOs such as Thames21 have worked with the Environment Agency, Lewisham Council, London, local trusts (River and Wetlands Community days) and networked 'friends of' groups via Facebook to be mobilized as *stewards of the river* in the face of local authority funding cuts. Meanwhile, my own river walks with *Talking Birds* theatre-in-place company who have created sound and spoken word performances; and, the Calotype photography with artist Jo Gane along the River Sherbourne in Coventry, UK, frame my approach to mediating the urban river.

Key questions raised by all of these projects despite their national contexts are: What do we already know about this river (i.e. it's history, industry, commercial use, meandering, flood and drought history, canalization, species, water quality, boating, swimming, fishing rights etc.) and what values have we had and do we have now about this river? What kinds of plans can be drawn up and developed about this river in collaboration with proximate communities and other stakeholders (i.e. the introduction of wildlife habitats, old and new species, water level management, new leisure experiences) and how can we bring these ideas to local planning governance, river maintenance, but how (if at all) does this influence policy (environmental, economic and cultural)?[13] The synergy of a worldwide movement of river clean-ups and de-canalization with creative, media and cultural heritage projects of public engagement has led my ongoing media and water research around five key research questions that have a strong arts and humanities approach:

---

[11] See also Paul Talling's *London's Lost Rivers* book (2011) and website. The Ravensbourne has benefitted from a strategic project (2010–15) including flood alleviation at Lewisham, citizen science water testing, river clean ups, wade walking, mind mapping and storytelling with London Bubble.

[12] Daylighting hidden waterways is now a worldwide endeavour, particularly in cities, where de-canalization goes hand-in-hand with new cultural expressions and memory, heritage work. I have seen some of this first hand in São Paulo Brazil, and the *Rios e Ruas* (Rivers and Streets) initiative project (see http://www.mostrarioseruas.com.br/) is an excellent example of a multi-media installation that has toured the city's cultural centres.

[13] For more on river restoration projects see Smith et al. (2014) 'The Changing Nature of River Restoration' and Tricia Cusack (2010) *Riverscapes and National identities*, as well as Schmidt and Mitchell (2013) 'Property and the right to water'.

- What is the role of narratives, memory, creative arts practices, community storytelling and archives for a variety of river stakeholders: such as city communities, local authorities, environmental agencies, urban planners, water companies and flood-risk managers?
- How can a trans-national memory approach to hidden city rivers and knowledge exchange enable community generated hidden river stories to re-engage people with a cultural memory of the social life of water?
- How do community-based and mediated storytelling projects around canalization, de-canalization and river culture sustain good water narratives that can be recycled as urban learning in a polluted, stagnant or repressed city river environment?
- What mechanisms can be produced for a transnational sharing of stories of hidden city rivers (through media) to sustain a cultural and collective memory of city-water ecosystems?
- How does this research help reverse the negative associations that communities generally have in relation to rivers flowing in their neighbourhoods?

Such questions resonate with a wide range of active river culture and heritage projects that are seeking to change perceptions of urban water. From the voluntary sector, grassroots media making and activism, social networking, wade walking and new creative industries, there is emerging out of both the UK and Brazil (two settings I have been most engaged with) new kinds of mediated water culture, history and developing media infrastructures. Seeking to engender a sense of community, civic engagement and citizenship through water storyworlds, such projects tackle head on how to develop an audience, how to brand and communicate water issues, the role of community cohesion, loss of community and community resilience around rivers, as Lejano et al. (2013a: 2) write:

> Stories or narratives create the glue that binds people together in networks, providing them with a sense of history, common ground, and future, thus enabling them to persist even in the context of resistance. It is through stories that people both analyse and realise personal relationships with land, animals, rivers, air, and even bacteria as well as the new technologies that impact the environment.

Towards riparian media and culture

Therefore, the emergence of what I will term 'riparian[14] media and culture' is based upon the importance of rivers in creating a specific cultural eco-system (particularly in the city) and it challenges the ownership model in Western/developed contexts

---

[14] The chapter draws on different understandings of a 'riparian' zone in research framing. UK Environment Agency publication *Living on the Edge* 2016 defines 'riparian landlords' as legally responsible land/property owners on the edge of rivers. 'Riparian' also refers to the interface of land/ water and the flora/fauna (biome) inhabiting these margins/corridors.

because it focuses on cultural memory rights and environmental responsibilities of the roles lived by people, who go to the river. Moreover, the *liminal or transitional zone* of 'riparian' functions as important and informative for engaging those at the 'margins' who 'immerse themselves in rivers', remember 'daylighted' water, forget canalized water and turn their backs on visible rivers. In terms of de-canalization projects this requires the use of media, cultural and creative producing to ask what role did and do hidden rivers play in lived daily lives? What kinds of cultures can be daylighted from swimming competitions, leisure pursuits, clubs, theatre, clothes washing, folk songs, puppetry and entertainment, photography, and bathing cultures in these urban rivers of the past? Thus, broadly how were these rivers considered and remembered as a source of and place of a social life and how can they be again in a post-industrial context without excessive gentrification that excludes ordinary expressions of a watery sense of place?[15] Such expressions can be found in the smallest of ways as can be seen in Figure 8.1, a photo taken during a scoping exercise of the river regeneration of Ladywell Fields, Lewisham, London, wherein the River Ravensbourne was decanalized and permitted to meander. One can just see the graffiti under the word 'Fields' which reads 'No Yuppies':

Figure 8.1 The new sign at Ladywell fields drawing attention the regeneration of the river (photo taken 2016. Courtesy of Andrew Holmes of the *Sustainable Flood Memory* team).

---

[15] In *The Politics of Urban Cultural Policy: Global Perspectives* edited by Carl Grodach and Daniel Silver (2013) one chapter (Lee and Anderson 2013: 74–5) explores the reopening of a stream in Seoul, South Korea, the Cheong-gye Cheon restoration project as part of the cultural vision of the Mayor. They report that the stream ran through the city centre, was paved over, had become run down and the stream was a sewer. By involving citizens in tours of the area, the Mayor gathered support for bringing back clear streams to Seoul, proving popular. The stream opened in 2005 and attracted thirty million visitors.

How, then, does the river re-emerge as memory and as flood even when expected to take a new course as an urban riverfront? What role do stories of good rivers, bad rivers and hidden rivers play in these city scapes? How are these polluted rivers (socially and physically) recuperated through creative, cultural and media practices? How are cycles of life, memory, waters, rivers and floods materialized in these cities?

We are, then, becoming aware of the importance of hidden rivers and streams in the city for creating a specific eco-system,[16] disrupting urban living during floods, and providing opportunities to remember and forget living with/without water. The aim would be to understand how we can use the research data collected on hidden river memory to understand the life-story of the river, its narrative and mnemonic practices (through a variety of media) for researching how communities experience their environment in terms of good, bad, emerging, erupting, corrupting, polluted, social and playful water courses.

Riparian media and culture directly address, create and circulate cultural values of water within-between urban riparian and water governance communities. A cultural and media policy approach to water can be effective if it produces shareable cultural messages for the benefit of riparian community with the aim to transform understanding of water in the city. As the Seoul case mentioned in the note above attests, it requires interdisciplinary research,[17] inter-professional working and a wide inclusion of diverse stakeholders and river dwellers to generate new frameworks for both cultural, media and memory studies, and water policy and practice. It emphasizes stories of human agency from engineered, natural-nurtured, 'daylighted' and forgotten water in cities and it does this through iterative and dynamic public engagement through concurrent *collection-recollection-circulation* phases where stories of the river's past, present and future are allowed to emerge. It is, quite simply, a social and cultural mediation of rivers at different stages of the nurture-neglect continuum by creating urban riparian community and memory in the first place, so as to then develop theories, policies and practices of water governance.

How might this work methodologically? How does 'memory research' and 'memory work' with river dwellers and users contribute to a better understanding of what urban riparian nurture-neglect actually means? How can media and creative production approaches (such as video diaries, social media content production, digital storytelling, photography, and oral history interviewing, developing river memory apps) support marginal and marginalized communities by facilitating the expression of their memories and aspirations into a shareable cultural message? To elucidate the capacity for memory to be 'actionable knowledge' and amplifiable in policymaking

---

[16] See Lavau's (2013) creative river narratives in order to explore sustainable water management of the Goulburn River, Australia, where in she sees irrigation water and environmental water intermingling but ontologically distinct through gatherings of practices, technologies and stories that constitute river management (Lavau 2013: 417–8).

[17] As noted in this book there have been a number of projects in the UK context that have taken this approach: *Care for the Future-Connected Communities* priorities of Research Council UK funding agencies can be found in 'Water in Future Cities: The Great Think', AHRC *Towards Hydro-citizenship*; ESRC *Sustainable Flood Memories*; AHRC *Multi-Story Water*; RCUK *DRY Project*; and AHRC *The Power and the Water*.

contexts, such an approach seeks to employ 'participatory' methods (my own have been inspired by the 'social memory technology' approach of the Museu da Pessoa, in Brazil, cited above). Yet these values need 'banking' for new cultural values of urban water become liquid assets to riparian communities and they need to be stored, saved and generate interest from new audiences. A dynamic memory reservoir of water stories for connecting cultures and identities is only possible if media and digital connectivity is embraced. This not only imbues urban rivers with mnemonic capacities, it reconnects people to discourses on *water memory* (chemical composition and somatic experience), *river memory* (community engagement and marginal/ized voices) and *liquid memory* (uneven flows of media and policy) in urban settings. Comparative analysis of urban riparian scenes can facilitate knowledge exchange pathways between contexts at different stages of river revitalization. Yet however inventive the methodology, it will have little impact if it does not change policy at the macro and meso levels and practice on the ground. Such methods must engage urban communities, organizational actors, policymakers and researchers in media, memory studies, cultural and urban geography, area studies, heritage and history, environmental management and water science and governance.

It is possible to trans-culturally 'story' water by 'constructing complex environmental information' (Lejano et al. 2013a) to address lesser-heard voices at the margins. Media reporting on rivers around the world and the online archiving of media stories are already making these connections, as noted at the beginning of this chapter. Yet we need to push the relationship between memory, lay/local knowledge and resilience (McEwen et al. 2012a, 2012b) into the trans-cultural if we are to connect the global circulation of water stories. While mental maps may address place-identity (Gould and White 1974; Lefebvre 1991), the affordances of recuperating, comparing and connecting urban riparian community memories remain largely unexplored. Therefore, linear weaving of stories (Ingold 2000) across contexts needs to be disentangled to circulate through a new spherical and transversal archaeology of intimacy (Sloterdijk 2011).

It is also possible to address riparian media and culture intergenerationally in that memories and stories of older residents can play a central role in sustaining an urban riparian sense of place, through an examination of mediating the 'nurtured' and 'neglected' river, drawing together conversations, oral history interviews, archival research, video and digital storytelling. Such storytelling activities should also remain alive to the status of women and the 'feminine' history or river cultures, as can often be found in the names themselves (such as Ladywell Fields, London or in the indigenous name for Rio das Velhas as Guaicuí, a derivation of 'guaimi', which means 'old lady'). Here the fertility-sanctity of water, folk rituals, water for ill and well-being and the erasing of quotidian practices such as washing clothes, singing ritual songs and sharing life stories can be unearthed. The key is to use media practices that ensure permeability to unlock stories through time-flows as well as via spatial intimacies. The storytelling spheres that these media practices should attend to are: (1) Water, as corporeal and somatic experience; (2) River, as physical-material determination of community; and (3) Liquid, as abstraction via media, data and story. Memories, collected on river walks and wades and from archives can be recollected through storytelling and media into digital stories and video shorts for social mediation and storing in an online memory resource, such as an app (as highlighted in Chapter 6) a local archive or through established media networks used by communities. Such a resource can be circulated to illuminate 'governance gaps' in urban water policy to explore what the memories of the Future River will look like. Such media content can be found in the media and creative practices of the UK NGO Groundwork, and in what remains of this chapter I would like to focus upon *Jordan's Story* (2016).[18]

Case study: 'I'm helping the river, but it has also helped me'

Jordan's Story (2016) is a two-minute and thirty-seven-second narration to sound effects, folk music and video, covering the single individual and small story of a young Welsh man reflecting on his teenage years. It has the strapline 'I'm helping the river but it has also helped me', setting the tone for a story about water, intimacy and mutual well-being. The fact that he is called Jordan and his story meanders with the river should not be lost on the viewer/listener. This is not a particularly groundbreaking story. It is Jordan's story. Jordan is not well known outside of his small Welsh community but his life is imaginable. He is young, reflective, remembering and through volunteering has improved his mental and physical well-being. He is (literally) in the river, undertaking the blue and green space clear-up work Groundwork is well-known for. Jordan is looking backward and forward at his own life (through memory) and at his environment (through the point of view of the video camera). He has had a traumatic past, but he is hopeful. This is an emotionally charged promotional video that uses life experience and memory to invite the viewer in. It delivers a watery, forested and mountainous sense of place. It chimes with Welsh tourism's media templates and new cultural narratives from natural resources policymakers in Wales. It may not circulate widely and may never reach (without a real push) its target viewers throughout Wales or beyond. Yet it remains Jordan's story and within it I want to locate two factors useful for riparian media and cultural studies.

- *A liquidity to memory* – by which I mean a form of slower cultural memory that has not taken flight from working within and with the economic and natural resources necessary to its meaning-making and spreadability. This story is highly mediatized, embedded in the cultural policymaking of Welsh narrative templates and promoting the natural resources of Wales as the place for health, exercise, well-being and non-urban everyday experience. It is a memory of the river in Wales replete with the cultural assets that speak of a watery sense of place as therapeutic, economic and spiritual.
- *A watery environmental activism* – by which I mean a form of mediatized hydro-citizenship directly targeting a younger audience that involves day-lighting the lived intimacy of river life stories, as well as the slow cultural labour of taking action within a care-ecology of human actors helping non-human actors to

---

[18] *Jordan's Story* (2016). Available at https://www.groundwork.org.uk/jordans-story and on YouTube at https://www.youtube.com/watch?time_continue=2&v=f2FoFJ4vPOk. Accessed 28 July 2018.

reshape their life worlds. This is a story about masculinity being re-thought in the framework of environmental care, compassion, solidarity and social movements.

Very typical of the thousands of 'digital storytelling-style' stories that are now very commonly made by NGO media and community engagement departments drawing on local footage, local people and usually a local musician, poet, filmmaker, illustrator, performer, animator or artist. These stories are present-ist, of and in the moment, seemingly extemporized, simple and easily refreshed. They build brand identity around volunteer and action groups. They do not seek to communicate science and rarely address the scientific or climate change agenda. The speaker does not always engage in big policy or climate politics or represent a wider community defined by national containers. That is what makes these stories leaky, permeable, malleable and spreadable, and they require social and professional networks to travel more widely.

I am most interested in the implications of such storytelling for riparian media and cultural studies, wherein new research may become more deeply interested in how such mediations effectively convey environmental and science communication. In the context of Twitter hashtags such as those used positively in #oceanoptimism or negatively as in #droughtshaming (mentioned previously in this book), to the many activities and movements of NGOs working in water management, there is a vital role for these mini-stories from riparian media and culture. The role that memory (cultural, collective, social and communicative) plays in the meaning of rivers and streams for emplaced communities and people, who may be disconnected from their watery sense of place, is critical. From building tools for crowdsourcing volunteer effort, to gaming flood disasters, to the creation of flashmobs, there are many new ways of activating 'hydro-citizenship' as a form of mediated environmental activism from the margins.

Jordan's Story raises key questions that afford a cultural memory policy to water that would concern the safeguarding of the role of story, narrative and anecdote in the management of river resources, the communication of environmental science and the cultural value of those resources. Such questions might be: Who gets to or is invited to speak about the environment and to whom? Who is listening? Who is brokering? Where is the public university (as space of science, art, social science, business and stakeholder)? Where is the media or memory scholar in this exchange to ensure that citizens are not simply mass data collectors: that their values, attitudes, life stories are integrated into the policymaking? Who gets to tell stories of water that improve its management (news media, the water companies, the environmental agencies, governments, academics, an emerging movement of hydro-citizens socially connective and active)? Again, what is the role of the academic, the policymaker, the governance structure here: to listen, collect, critique, nudge, chastise or to be the active and honest broker in the exchange and circulation of cultural values of the river? How can media practitioners nudge and tickle memories to the surface and what is the professional and researcher's responsibility to these stories after they have been told?

Jordan's Story is very similar then to the many stories I have been a part of collecting and co-producing with a wide range of stakeholders in the two funded projects I have mentioned in previous chapters: the ESRC funded *Sustainable Flood Memories Project* and the RCUK funded *DRY: Drought Risk and You Project*. As well as in follow on funded projects such as the 2018 *Re-connecting Children with Water Memories* (with Peter Kraftl, Birmingham) and the 2019 project *To Walk with Water* (with *Talking Birds* and the Centre for Agroecology, Water and Resilience, Coventry). The key features of these projects are their *strong inter-disciplinarity* between the arts and the sciences offering a science-narrative integrative approach to the research and media production. *Memory activation* – remembering as a creative act and social contract between rememberers, researchers and water stakeholders (businesses, water companies, government, health services, cultural organizations and agriculture, for example). Here cultural memory research (on trauma, witnessing, resilience, deep listening and imagined futures) are central to expressing the messiness that is the human experience of environmental awareness and living in distress. *Media discourses, forms and practices* (photographs, video, social media, animation, digital storytelling, memes and GIFs, for example) as ways of expressing difficult pasts, uncertain presents and imagined futures. Daylighting hidden narratives, creating, sharing and circulating these mediated memories of a dry and watery sense of place, and spreading memories as a personal storying of water.

Jordan's story, while not directly produced by our research project team was created by Groundwork, one our partners in the Wales case study (one of seven) of the DRY project (cited above), during our time working with them. It is an NGO established in 1990 operating across Wales, mostly in the post-industrial mining regions, and its strapline is *Changing Places, Changing Lives – One Green Step at a Time*.[19] 'Wales' is iconic in climate-related narration. It is the benchmark for sizing up any global natural disaster in the British imagination, a shorthand phrase used by journalists to proportion the impact of earthquakes, floods, drought, forest destruction, iceberg melts, for example, all around the world. 'An area the size of Wales'[20] has become common parlance just as in Denmark one might say an 'area the size of Fyn' [størrelsen af Fyn] to determine in the audience's mind how large the issue being mediatized is. This association of natural disaster with an area the size of Wales has become a part of the national and cultural imaginary of Wales itself and has even led a rainforest charity to brand itself the *Size of Wales*, to create a platform for forest charities to signal their collective efforts to those inside and outside Wales (see Size of Wales [Maint Cymru] at https://sizeofwales.org.uk/).

Wales itself is surrounded on three sides by the Irish Sea, on the fourth side it is bordered by mountains, the ancient Forest of Dean and two main Rivers (The Wye and the Severn). Within its territory, it is rich in woodland (particularly coniferous) and with many river catchments, making for a 'wet and green experience', fast (or flashy) run-off of water, numerous reservoirs, waterfalls and a 'commons landscape'. It is also socially and economically deprived compared to England with European Union funding used to regenerate many parts of post-industrial South Wales, after the decline of the coal mining industry. Forgetting the trauma of the past has been Jordan's personal strategy while lost in his gaming console, and the collective strategy has seen South Wales (where he lives) produce the lower-wage, highly skilled labour

---

[19] I have researched campaign strategies with the CEO and Head of Communications of Groundwork on their communication campaigns during 2018–19.

[20] Wikipedia lists it under 'unusual measurements'.

that the massive Sony UK Technology Centre needs, the main factory sited there.[21] A combination of well-remembered mining disasters, chronic lung disease and physical scarring of the landscape mean that any nostalgic living memory of the carbon-based energy extraction of the region is rarely encountered. In the last ten years, significant and co-ordinated cultural and media policy has changed the *storying of Wales*: to be remembered now as the carrier of environmental and regenerative messages. As a place to '#Find your Epic' and the 'Year of Legends', we find that Welsh water is taking on a new agency, as in the 2016 and 2017 television adverts popular on YouTube, in which the Welsh actor Luke Evans heads up a computer-generated image of a wave rolling up behind him across a lake/reservoir to promote Welsh Tourism.

The Natural Resources Policy has given Natural Resources Wales (2016–17) real teeth based on the *State of Natural Resources Report*, where 'economic, social and cultural benefits' are given equal weighting. Culture is not the poor cousin here to commerce. As the main supplier of water to England; the key producer of renewable energy from off coast and inland wind farms; a supplier of wood, cork and paper; it is also the outdoor pursuits region of the UK, with coastal paths, mountaineering, military training and many heritage and historical sites. With that watery sense of place comes a politics of identity that sees Welsh water as belonging to Wales, which during drought causes significant perceptual grievances towards the English (see Morris, 'Soak the English: Welsh want paying for any water piped across the border' *The Guardian* 29 February 2012).

Therefore, if we are to address perceptions of water at national and local levels then cultural and media policy studies of memory, heritage and social movements should be embedded in economic resources management such that we approach rivers, in particular, culturally, socially and creatively. As noted in the Introduction to this book, Bell and Oakley's *Cultural Policy* (2015) calls for cultural policy studies to move in new directions such that culture is not seen only as the context in which policy acts but that culture acts on policy. Thus, water and rivers need to be researched as meaningful to people and 'valued in terms other than the economic – or through a radical rewriting of the definition of the economic' (2015: 157–8).

Jordan's Story is not then the first or only story about water or activism, climate change or future scenarios. It is not a channelled story, going in one direction from one source through one resource to a reservoir of stories to be used up later. It is, in fact, day-lighting a young man's intimate experience with environment, with a care ecology at the centre of the narrative: *for if I care for this one thing it helps me care for other things* it seems to say, and it does this through remembering. Like the river, Jordan's story meanders, and we follow it as it winds through personal experience, first-person narrative, reflective of self and through the river as reflective surface. He is immersed in his iterative life, he climbs high up to gain some perspective, out of the water and the woods he walks up to the mountain tops where precipitation, run off and snowmelt act as the source of the rivers below. He gets into the dirty and junk filled river to see his environment from the river's perspective, and he cleans it up with others like him but not like him. He does what video makers of river wade walks have been doing for some time: *Jordan sees what he thinks the river sees and feels and he acts.*

---

[21] It should come as no surprise to memory researchers that this is where Sony also operates its Media Lifecycle Solutions, where they aim to give archive media a new lease of life.

Unlike most media stories about water and environmental activism this is not alarmist, apocalyptic or regretful of a failure to act, neither is it anticipatory of future disaster, and it does not address macro-scale issues of flood, drought, water management, the party politics of climate change or past practices of industrial pollution. It neither works within dominant media representations (the six framings of climate change as Mike Hulme calls them)[22] nor does it work with dominant scientific explanations from hydrologists or water scientists. It creates a watery sense of place as deeply connected to mental, emotional and physical well-being. It does not de-centre the human but neither does it decentre the non-human. It centres on personal action, for a deeper search for social, cultural and personal values by not discounting the human altogether and by championing deep and slow remembering as a form of anticipatory memory (see De Silvey et al. 2011 on 'anticipatory history').

If we then take a cultural and media policy studies perspective to the emergence of a riparian media and culture, we must address the increasing human and creative labour involved in a new kind of mining of data by humans of those non-human actors (rivers, trees, floods, droughts and heatwave). Alongside calls for ramping up citizen science there should be recognition that people are more than simply data collectors. There is a growing demand for artistic, cultural, craft and creative approaches to environmental citizenship and science communication within many research and NGO projects. The arts and humanities is being asked to press the 'call to action' button, but how they respond and with what kinds of ideas about human and non-human relationships cannot de-humanize the storytelling.

There is also a new focus on children and young people as increasingly entangled and responsibilized for parental and generational failures, and thus they are also called to action, to repair, to regenerate but not (necessarily) to regret. Jordan exemplifies this in his story of the river who helps him and how he repays that help. This is not responsibility this is repayment for a value that has been added to his life in a new liquidity of well-being. He emerges (alongside changes to the UK's geography curriculum) as an actor and agent, with a new experiential lay knowledge of the river, offering a liquidity to his memories that have widely shareable values. Young people now are seen as 'young people in the future' not the present, they are being responsibilized 'now' to be better consumers going forward, to be better rememberers. Much of this may involve nostalgia and deeply classed and gendered assumptions about young people spending more time outdoors 'then' (whenever then is chosen to be) than 'now'.

However, re-connecting young people and childhood more broadly with cultural memories of the environment, in which the young emerge as compassionate actors at various points in time will be an important intervention from media and memory studies. Thus, I propose that water's storying be visualized as bubbles in close connection to one another (see Figure 8.2) in which theories, methods and practices for riparian media and culture coalesce around water memory, river memory and liquid memory as three distinct but intimate ontologies imagining the pasts and futures of rivers.

---

22 See Mike Hulme's very comprehensive website of his approach to framing weather and climate change, where he covers new research on newspapers, https://mikehulme.org/and his book *Weathered: Cultures of Climate* (2016).

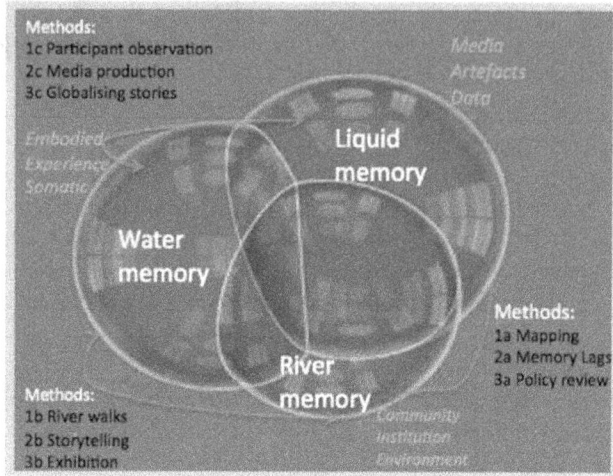

Figure 8.2 A Riparian Media and Culture Bubbles Method for civic engagement in river research.

By making transparent media, memory, water, river and liquidity in the context of mapping, participation and policy, the methodology seeks to address the following key questions for future river regeneration research:

1. What stories get lost-forgotten and found-learned in river neglect and regeneration and how can these be mediated, shared and exchanged?
2. How are those who 'remember' the urban river transformed through their immersion in its stories and memories?
3. How mutual is the feeling of an abundant or degraded sense of place across river-dwelling communities, and how far can they express their riparian aspirations in a cultural message?
4. If stories, narratives and memories can reinvigorate community engagement with (urban) water, as is often said, then how do we understand and evaluate the marginal, peripheral, hidden and lost memories of the river in producing new cultural meanings of water in the city?

Therefore, if as Sloterdijk argues in *Spheres: Volume 1 Bubbles* (2011: 10) that 'effective philosophy begins by splitting society into those who remember and those who do not – and, furthermore, into those who remember a particular thing and those who remember something else' then we can see that memory work will be crucial to the success of mediating water for future generations.

9

Waterproofing media and memory for flood risk

Introduction

The headlong stream is termed violent
But the riverbed hemming it in is
Termed violent by no one.

Berthold Brecht 'On Violence' *Poems 1913–1956* (2007)

It may make just as much sense to curb our compulsive, single-minded efforts to control water through elaborate structural interventions of steel, bricks, cement and embankments, as it would to adapt to flooding through smarter use of data, science and early warning systems. The challenges we face now and in the coming years may drive us back towards embracing the previously dynamic relations between land, water and communities (Building Futures 2007: 5) as much as forward towards new uses of sensors, digital technology and citizen observatories.[1] We know that floods can be catastrophic. They are mediatized nationally and globally as human stories of natural disaster (see Chapter 2) that address a politics of vulnerability of humans to their environment. In popular writing such as John Withington's (2013) *Flood: Nature and Culture*, he converges Noah, the China floods of 1931, the broken levees of New Orleans and the flooding in Europe into a global historical and mediated repertoire of deluge (see Kempe 2003 on storying floods in Early Modern England). It is not surprising that floods engender such a long and generalist view of their iconicity and connectedness to deep time events considering their impact on communities and lives.

Yet flood risk and flood events appear to demand and require a rational multi-stakeholder response, debated in the public sphere and enacted through Centres of Expertise and Civil Response. In turn, this might imply that flood agents and resilience actors behave rationally. Yet the stories and images of flooding are loaded with aesthetic, emotive and performative power. In fact risk communities engage in forms of flood heritage that their community may be built upon or a flood-prone culture may be increasingly mediatized by new archives of social networking, such that there is an emerging environmental literacy often ignored by flood-risk managers.

---

[1] See Degrossi et al. (2014) 'Flood Citizen Observatory: A Crowdsourcing-based Approach for Flood Risk Management in Brazil' in 26th *International Conference on Software Engineering and Knowledge Engineering.*

This raises the first key problem: flood risk is addressed top-down, on a context-by-context basis, and communicates dependency. Rivers, for example, may flow, through urban communities, rural regions, watery nations and evoke 'wet' identities but rivers are inherently 'trans' and memories of them are mobile. The opportunities for social learning is across river catchment regions. Flood knowledge needs to build on notions of folk or indigenous memories of flooding, as well as lay expertise and social learning by seeking strategies to inter-generationally communicate with and to audiences using new tools. Therefore, to what extent does flood memory open up possibilities for future resilience? Moreover, how do these communities continue to record and preserve their environment's memorable changes?

For communities to be aware of flooding (and very importantly to prepare and take action) as a form of 'socio-ecological resilience' (Adger et al. 2005: 1036), then they need to record and remember in some (everyday) way. Thus, it is very difficult to ensure policies compel people to take any action without them remembering prior flood experience (either personally, collectively, virtually or prosthetically). Even if communities have experienced flooding, as my research has so far discovered, they not only need *repeat experience/anecdote* but they need to deal with the social compulsion to forget in the light of the mediatization of flooding as depleting, disabling and destructive of their homes and communities. Memory, then, becomes a *way in* to resilience planning. Moreover, the building over the flood plain has hidden many of the landscape markers former generations used to prepare: *when that field is flooded, I know I have one day to clear the downstairs*, was a frequent early warning measure mentioned by experienced flood plain dwellers in parts of Gloucester. In fact, in a relatively deprived part of Gloucester there is a small area on the outskirts of a traditionally working-class community. It has an established core of elderly residents, alongside a transient younger socially housed population, as well as a settled Gypsy, Roma, Traveller encampment. In this community there is one man many turn to for their flood memory. Known locally as Noah he can tell you anything and everything about the River Severn.

More useful to local people than the Environment Agency, Noah's memories and lay knowledge of flooding in this area developed over seventy years of living in the same location, have been documented in his own archives, shared online and recorded and uploaded to YouTube. When it rains, and there are flood warnings, the locals ask Noah whether to take action. We interviewed Noah many times. The relationship between flood memory and personhood is not static, as the geological or geographical metaphors of memory would suggest. Noah does not offer strata or layers or easily mappable data. Neither as Hillary Mantel suggests are 'all the memories laid side by side, at the same depth, like seeds under the soil' (*Giving up the Ghost a Memoir* 2003), rather flood memory is just waiting for memory researchers to uncover them. Thus,

> memory has a living, shifting timespace dynamic, perhaps formed of living, convulsing, labyrinthine entanglements of many virtual (living) landscapes in unconscious memory stores, which are often dark (as in not illuminated by conscious recollection) and thus unmapped and unmappable. They are subject to the fluidity of emotional/affective convection currents (and more violent disturbances) as we live on.

> (Jones and Garde-Hansen 2012: 13)

Thus, as suggested by Jones (2011: 8) memories, in the context of flood research, 'are living landscapes seen obliquely and from an always-moving viewpoint of ongoing life'. They are the powerful economic drivers to forget and ignore the movement of millions of gallons of water through our built and farmed environment that require our commitment to mapping the unmappable through closer and more intimate participation with flood-risk communities.

Towards participatory flood cultures

Participatory methods of research and engagement are not new. In the visual arts and media, such methods give the means of representation over to participants, permitting, as in the memory work of previous chapters, communities to 'frame their *own* lives, tell their *own* stories, represent their *own* situation, offer their *own* understandings'. So argues Maria Pini (2001) on her use of video diaries by participants for social science research. 'Narrative approaches' in citizen science, say Constant and Roberts, 'allow for unique, context-based evaluations through time-oriented structures (event-focused, causal, temporal) revealing how changes occur and evolve from a personal perspective' (2017: 4).[2] They are not without their challenges, least of all that there is little acknowledgement of creativity, subcultural and subaltern activity, experimentation and the opportunity to seize the means of cultural production and shape the narratives in ways that do not fit the research or policy agenda. As Jon Coaffee has argued 'the building of resilience will be most effective when it involves a mutual and accountable network of civic institutions, agencies and individual citizens working in partnership towards common goals within a common strategy' (Coaffee et al. 2008: 3).

As science communication or the 'public understanding of science' has been an explicit goal of current research agendas in the water sciences and the social science domains concerned with floods, drought and river management, alternative methods for brokering knowledge have proliferated. It is media (their devices, networking, speed, connectivity and convergence) that have democratized science communication in ways that has moved that agenda closer to civic engagement and across national boundaries. From a *knowledge deficit model*, in which the goal is better information in a one-way flow, to more collaborative and co-productive understandings that seek to flatten hierarchies, challenge concepts of expertise and afford diverse forms of knowledge, from the ground up. Implicitly, these participatory approaches have been afforded by the growth and technical developments across (often global) media, cultural and creative industries, and are informed by the genre, narrative and textual methods of analysis used by media professionals and media researchers.

Thus, the idea is that the scientists and the public co-produce knowledges of water through engagement with an attention to context, demographics, place, values and settings. These all sound rational enough but in my media-based experience of

---

[2] See also Keightley, Pickering and Allett (2012). 'The Self-interview: A New Method in Social Science Research' *International Journal of Social Research Methodology* 15: 6, 507–21. Keightley and Pickering go on to frame the media and memory research domain with their later publications.

brokering collaborations between communities and researchers on water topics much gets forgotten or remains unspoken about the cultures that citizens practice every day. Not least the moral, social and political values, and the trust, empathy, spirituality, personal interests, tastes, memories and preferences, many of which offer alternative forms of knowledge and culture; and, are highly mediatized and increasingly socially mediated. While there may be confirmation bias wherein participants reinforce existing beliefs, there are also opportunities to create different spaces and places for water communication, ones that are media-based, creative, safe and experimental, that can offer opportunities for new knowledges to be co-created.[3] Such an approach needs the arts and humanities, arts and cultural partners leading the communication and creative methods of capturing and analysing *stories as data*, and the digital creativity of participants for iterability and interactivity. This combination of experience, knowledge and skills is the challenge, alongside audience development strategies, branding and marketing techniques, avoidance of participant fatigue or the danger of only attracting already well-served publics. All of which requires recognition of the importance of memory, anecdote, values and narrative (key to everyday life) and that scenario-ing is not restricted to the science domain. As Paschen and Ison (2014) argue, on narrative research in climate change adaptation, and March et al. (2012) make clear, on 'water futures' and water-scenario analyses, scenarios are important in times of uncertainty. They allow for plurality, complexity, iterability, affect, emotion and dynamism. Yet few studies have engaged with arts and humanities (and media studies in particular) for an understanding of storylines, narrative, protagonists and what the idea of scenarios as 'memories in the future' might mean in practice (Weick 1995: 185). As Burnham-Fink (2015: 48) claims:

> Scenarios are stories. In the diverse field of scenario planning, this is perhaps the single point of universal agreement. Yet if scenarios are stories, their literary qualities are often underdeveloped. Scenarios used in business and government frequently do not contain a relatable protagonist, move a plot toward resolution, or compellingly use metaphor, imagery, or other emotionally persuasive techniques of literature. In these cases, narrative is relegated to an adjunct role of summarizing the final results of the workshop. While this neglect of narrative may be reasonable in some contexts, the power of narrative should not be underestimated.

As I have noted in a previous publication on flood-risk scenario-ing: '[f]rom local modes of storytelling to national news and social media networking, floods are iterative, itinerant and relational, offering up insights into change as well as acknowledging continuities in local community experience' (Garde-Hansen et al. 2017: 385).

Consequently, in this final chapter in the *Perception* section of this book, I wish to foreground emerging research collaborations across national containers of water management that draw into their thinking, methods and practices, global media

---

[3] Quigley and Buck (2012) propose a 'third space' through Photovoice where scientific discourse and everyday discourse are authentically integrated toward problem solving.

and digital culture for connecting people to mitigation and adaptation strategies. By focusing upon flood-risk management using media, memory and data for scenario-ing, I am addressing that flooding especially in cities has the biggest impact and is varied in its causes (pluvial, fluvial, coastal and flashy), with diverse results. From inconvenient disruptions to the flow of transport to mass casualties, from the benefits of seasonal floods for farmers to new opportunities for capturing water to mitigate drought. It is also the water event that has been the most mediatized (as shown in Chapter 2), benefits from a wealth of archival, heritage and cultural memory materialization (as explored in Chapter 6), and is built into the key dynamics of *remembering-forgetting*, as communities grapple with how to learn from past floods while wishing to *move forward*.

Sensing floods: Data as the new levee?

In Chapter 6, I considered the concept of flood memory as metadata by drawing upon the materialization, memorialization and mediatization of past floods in environments. Thus, we were focussing on the cultural and collective memory of flooding in particular places and settings and communities, and flood memory afforded a narrative framework for storing, organizing, building, retrieving and circulating flood stories. However, at the micro level we have flood stories as data and here a more logical and scientific communication of flood may be expected, ones that can be seen played out in public spaces, wherein clouds, data, smart technology and scientific analysis are globally advertised as the principle matrix through which flood risk will be mitigated. For example, in the photograph below (Figure 9.1) the digital banner at London Heathrow Airport's Departures area, above the security clearance, advertises Hiscox's collaboration with Microsoft on a data-led approach to flood risk as one of its 'Azure Stories' which cycle through the banner while passengers wait to pass to airside. In twenty words and two numbers we learn that two large tech companies are solving climate-related risk with data and speed.

Yet, how is this *a story*, and how can flood-risk communities address such a technological response to water risks, without the inclusion of human stories and stories of humanity, to ensure a more stretchy understanding of 'data'? If data's 'purpose is to be generalisable' and work 'within a paradigm where it is possible to represent a universal truth in an objective fashion' then how can data also be stories which are 'specific, context dependent and work as examples'; for in 'a narrativist paradigm, all knowledge is understood to be partial and truths are subjective. Narratives have a different type of legitimacy and function' (Constant and Roberts 2017: 5). The danger of *levee syndrome*[4] continues if we only rely on this narrative conceptualization of fast water data to hold back the water.

---

[4] We discussed our evidence of this during the Sustainable Flood Memories project. See our Wordpress site https://floodmemories.wordpress.com/category/levee-syndrome/. 'Levee syndrome' describes a condition in which the presence of safety measures decreases risk awareness and leads to a lack of preparation and a liberal attitude towards the hazard. See definition in *Disaster Resilience: An Integrated Approach* by Paton and Johnston (2006: 111).

Figure 9.1 The Departures entrance of London Heathrow Airport, November 2018. Digital advertisement for Hiscox above the security gate. Photograph taken by author on way to Brazil for the *Waterproofing Data* Project (2018–20).

It may make sense to see digital media, networked creativity, 'hackable' innovations as all expressing an upsurge in 'disruptive technologies' (which I will focus on in the conclusion): not unlike disruptive water events in their representation. These, not coincidentally in the light of the synergies of media and water explored throughout this book, were originally described by Bower and Christensen (1995) in the business sector in 'Disruptive Technologies: Catching the Wave'[5] as truly threatening to the status quo as well as sustaining. If organizations *scenario-ed* well, were just ahead of consumers, the public, users and audiences, and imagined alternative futures by drawing on memories, then they were *waterproofed* from technological change. But they needed to forget and even kill off some of their old ways of doing things.

In this context, flood story as data can be disruptive or sustaining if researchers and communities work together to define the strategic significance of the digital technology or creative media application for developing new kinds of flood memory and narrative for community resilience. Moreover, they will need to kill off two key ways of doing things in research and development traditions. Disciplinary territorialism and singular knowledge expertise assigned to one individual will need to give way to strong interdisciplinarity, collaborative research and interprofessional working.

---

[5] It is important to highlight that in the original January–February 1995 print copy of the *Harvard Business Review* in which Bower and Christensen's article appears, they finish their argument with a cartoon by H. Martin, showing a CEO at a podium saying 'Good Evening Ladies and Gentlemen' and above his head is the strapline 'The Night of the Big Storm'.

There is potential to converge social media, cultural memory, communication and flood data with the internet of things,[6] which Gubbi et al. (2013: 1645, my emphasis) have described as 'the next *wave* in the era of computing', wherein mobile devices or move-able sensors could be attached to river banks, reservoirs and within the built environment to allow the harvesting of data.

> This results in the generation of enormous amounts of data which have to be stored, processed and presented in a seamless, efficient, and easily interpretable form. [...] However, in the past decade, the definition has been more inclusive covering wide range of applications like healthcare, utilities, transport, etc. Although the definition of 'Things' has changed as technology evolved, the main goal of making a computer sense information without the aid of human intervention remains the same. A radical evolution of the current Internet into a Network of interconnected objects that not only harvests information from the environment (sensing) and interacts with the physical world (actuation/command/control), but also uses existing Internet standards to provide services for information transfer, analytics, applications, and communications.
>
> (Gubbi et al. 2013: 1646)

It is the phrase *without the aid of human intervention* that will require strong interdisciplinarity with arts and humanities approaches if media, creative and digital technologies are to be engaged in the story-ing of water, particularly if that water is a river (a non-human agency) with rights to memory, and the data produced is not designed for instrumentality, surveillance and control.

Extending the discussion in Chapter 3, then, we find that social media such as Facebook, Twitter, Weibo and Instagram can become deeply connected to these sensors and provide wider discussion, debate and decision-making (a potential for a *digital-hydro-public sphere* perhaps) for holding water managers and environmental agencies (and their governments) to account (see Moulaert et al. 2010). Such water sensing can work for a wide array of interests, not only commercial. Social movements are made possible when flood stories become connected across river catchments, basins, domains and flood-risk locations through tools for crowdsourcing volunteers (see previous chapter on Groundwork), through rapid response social movements (such as through Twitter, flashmobs or Ushahidi) and through creative media technologies for social learning (as in gamification, theatre-in-place or motion capture performances i.e. literally 'sensing the city'). All of this have the capacity to ensure that water, flood, rivers and drought are recorded, remembered, recognized and paid attention to, that is, cared about, especially where there are hidden rivers in the built environment. Thus, turning the one-to-many flood-risk communication of the twentieth century into a digital remembrance of all things water related, as ideas exchange, new communities form with the potentiality for collective action. If flood-risk communication tends to focus on wider global issues of climate change communication rather than localized

---

[6] The term internet of things was first coined by Kevin Ashton in 1999 to describe changes in supply chain management.

perceptions of water as the starting point for care of the watery environment (see Whitmarsh 2008 and Taylor et al. 2014), which is to be expected considering the media framings,[7] then how then can we make room for water in our daily lives? How can these perceptions become attuned to the water beneath our feet and the hidden memories of water in our cultures? Moreover, how can we ensure that all this digital remembering of flood-risk, flood events, duration and recovery allows for 'digital forgetting' and the 'right to forget' (see Mayer-Schönberger 2011) one's personal (and traumatic) relationship to the flood event?

Making room for water

Perhaps we need to do more to curb our compulsive, single-minded efforts to control water through elaborate structural interventions, move away from bricks and mortar-based solutions. The challenges we face now and in the coming years may drive us back towards embracing the previously dynamic relations between land, water and communities.

(Building Futures 2007: 5)

While the Netherlands rolls out its *Room for the River/Ruimte voor der Rivier* strategy (2011–ongoing)[8] and the UK seeks to convey a participatory message within a hierarchical governance structure (See the 2018 blog 'Creating a Better Place: Engineers and the Environment – natural allies in our fight for a more flood resilient nation'), there are emerging communication conflicts on climate-related risks. In these contexts, flooding (particularly in urban contexts) is less about adaptation and more about survival and acceptance.[9] On the one hand, various nations have either experienced a retreat *of* the state in flood-risk management (such as in the UK after the Pitt Review 2008, which called for increased civil society responsibility) while other nations (such as Brazil) have only recently begun to develop a more coordinated state/municipal response to flood disasters, even when flood-risk communities themselves retreat *from* that state. Researchers have noted that mediated flood narratives tend to connect human interest stories within wider discourses of class, politics, identity and society (see Furedi 2007; Murthy and Longwell 2013 and Escobar and Demeritt 2014). While some studies have focused on an individual's trauma such as the Hull Flood Project which used recovery diaries to capture reflections on health, social networks and economic well-being (see Medd et al. 2015) other studies on mediating floods have noted how journalism uses memories to reflect the future and not the past (see Trümper and Neverla 2013). Thus, in the Hamburg flood context 'disaster

---

[7] See Shukman (2014) 'Barrage over climate change link to floods' BBC *News On-line*. Available at: http://www.bbc.co.uk/news/science-environment-26242253. Accessed 5 March 2014.
[8] See the *Room for the River in Three Minutes*, 19 April 2016 video on YouTube.
[9] There are other ways of thinking emerging about the increased and necessary acceptance of intimacy of human spaces/places and waterworlds such as 'sponge cities', 'slow the flow', 'natural flood management', 'blue-green infrastructures' and more recently in policy terms the UK's Bricks and Water policy discussion of 2020 regarding building flood-resilient homes.

memory became a veritable duty' to engage with a mediatized 'memory landscape' of speeches, memorials, high water marks, signs, newspaper articles, photographs and films to offer 'a perspective for the future', as a coping and warning mechanism, by which 'commemoration events and medial or material acts of remembrance make up an integral part of local and national memory culture' (Mauch 2012: np).

In spite of the confusion around synergizing climate change and flooding, it is clear that '[h]earing others' experience or examples, sharing stories and fostering networks' are 'emphasized as ways to learn and adapt' (Bohensky and Leitch 2013: 485). This chimes with Haughton et al. (2015) who found that both expert scientific knowledge and local knowledge to be contested and contestable concepts, thus avoiding essentialist notions that one is more credible than the other and opening up the concept of 'lost knowledge' of flooding as inclusive of a wide range of actors (Haughton et al. 2015: 384). How could sensors, smart technology and flood story as data capture store, process and recirculate this lost knowledge is one area of research I am currently working on with a wide range of scholars and communities in the UK and Brazil.

Alongside climbing insurance costs for property owners in well-developed contexts, disengaged landowners and property developers or precarious housing in flood-prone areas (where there is no governmental or economic safety net), participatory approaches to flooding are increasingly complicated, and take a great deal of time, resources and attentiveness to cultural diversity, inclusion/exclusion, discourse and media technology access. Where, we might ask, are the people, citizens, communities, audiences, service users, taxpayers (or whatever other concept of 'the person' one wishes to use) in this global debate with its local needs for adaptation and risk management strategies? Whose voices and how are the voices articulated and what are they saying? How are different digital platforms, cultural projects, sensed spaces and places and social mediations being employed and through what kinds of communication practices by a range of sectors and organizational actors? What mediated connections of the scales of person to water to state to global are made possible? How could research from digital media and communication studies (such as around participatory cultures, digital memory studies, data cultures and storytelling) be used to explore and inform these debates and adaptations? Where better in the world, considering changes to political context around climate change discourses, to address these questions than in Brazil where I have researched since 2012 on media, memory and water projects with a range of water stakeholders. Which does not mean focusing on one media form, concerning one event in one context, as is the case for Bohensky and Leitch's (2013) social sciences approach to flood media; who in 'Framing the flood: a media analysis of themes of resilience in the 2011 Brisbane flood' used news frames to consider the encoding of meaning into newspaper articles. Finding an equal quantity of articles from their sample that offer dominant narratives of linking climate change to flooding as narratives that deny climate change, they shift their attention to discourses of community:

> Ideas of community spirit, cohesion and coping – all considered to contribute to resilience in theory – were reiterated in the media discourse. This was likely to reinforce the notion among both the affected and wider community that the city

was able to self-organize through agency, communication and cooperation, rather than promote the usual media focus on stories of helplessness.

(Bohensky and Leitch 2013: 485)

Thus, a closer focus on social and digital media in its community context and how it operates in place-specific ways gives clues to how audiences use media on the ground as integrated with their wider personal, professional, social and cultural experiences.

Transnational media waterworlds

In the areas I have undertaken media research (the UK and Brazil), floods are marked materially in the environment and through analogue media as a mixed *emo-scape* of national event, personal loss, trauma, fear, excitement, community solidarity, financial crisis, economic opportunity, collective helplessness, wartime resilience and even social and personal cleansing. Material marking of floods and their impact are a form of metadata (as noted in Chapter 6) that is 'structural', through flood marks, flood gauges, flood levels on buildings, at bridges, churches, in the landscape and that of the catchment itself. All of these can be both mediatized socially and sensed by internet-enabled technologies. These material and increasingly digital memories of flooding are both formal and informal, and speak to a 'natural history of [water] memory'.[10] Informally, river catchment communities in the UK and Brazil have marked flooding through plaques, photographs on display, in private albums, in newspapers and through personal mementos/memorials, as well as through immaterial storytelling, anecdote and performative modes. Such well-established forms of marking flood memory in communities with a watery sense of place may not, though, engage young people whose awareness of flood-risk water issues is increasingly mediated through global narratives and local social networks. They forget the very water beneath their feet. New technologies for marking and remembering flooding can move flood memory to online and digital spaces and places (i.e. animation,[11] digital artwork, photo-sharing, blogs, social networks, mobile apps, online videos,[12] games and tweets) and this has the potential to define and structure flood memory into new forms of

---

[10] See Bond, Rapson and Crownshaw's 2015 Project, *The Natural History of Memory.* Available at: https://naturalhistoryofmemory.wordpress.com/. They say of their project: 'While memory studies typically positions historical sites and landscapes as the places where past catastrophes unfolded, this project understands these environments as the very media through which these disasters took place, lent agency and co-opted by the perpetrators of those events, thereby enabling their occurrence.'

[11] See the animated data visualization of Bangladesh's Brahmaputra River over a thirty year dynamic history 'Scientists with the European Commission's Joint Research Center in Ispra, Italy, working with Google engineers, have used millions of satellite images to illustrate how rivers, lakes and other bodies of water have changed over three decades.' Available at: https://www.nytimes.com/interactive/2016/12/09/science/mapping-three-decades-of-global-water-change.html?smid=tw-share&_r=1. Accessed 14 October 2016.

[12] See Hicks et al. (2017) 'Risk Communication Films: Process, Product and Potential for Improving Preparedness and Behaviour Change'.

metadata. This, in turn, has social learning applicability that researchers and water managers should embrace.

Improving the resilience of cities to flooding has been recognized as a Sustainable Development Goal (SDG) 11 of the United Nations (targets 11.5 and 11.B). The *Sendai Framework* (United Nations Office for Disaster Risk Reduction 2015) which promotes a 'gender, age, disability and cultural perspective in all policies and practices' (5)[13] assumes that data[14] will play a prominent role. However, this goal makes an assumption that we all know what flood-risk data is and looks like, how it is produced, what practices take place in its application and that it is complete if we have more of it and effective if it moves fast. In Brazil, and in São Paulo in particular, floods are frequent (it is a tropical location) and have had significant impacts (it is a city built on top of its water and is mostly made of concrete). The Brazil Geological Survey (CPRM) has noted that severe landslides and floods, water-related disasters and protecting the poor and people in vulnerable situations are the key foci of flood risk in Brazil. How could flood storytelling and digital flood memory possibly mitigate this kind of disaster?

A sustainable solution to floods and flood risk requires adaptive and preventive approaches that include stakeholders and communities (United Nations, 2015). From the mid-1990s UK flood policy shifted towards a more devolved risk management approach, with increased recognition of the value of local knowledge and strategies but with less heard voices in community capital and social learning for adaptive resilience (Cabinet Office 2008). This value was still determined by economic definitions. Bound up within the concepts of 'resilience' is the ability to continue to function under stress and to bounce back to a better state of recovery, as well as placing resilience at the level of 'community' ties in local decision-making (e.g. the UK's Localism Act, 2011), with devolved flood-risk management in many countries.[15] Does this mean that lacking resilience to flooding is because a community not only lacks flood knowledge but lacks flood memory and is bereft of a sense of remembering water with few constructive ways of enculturating a watery sense of place? If the democratic governance processes of participatory decision-making in Brazil, for example, have been shown to strengthen resilience and increase the flow of information,[16] enhancing awareness and stimulating the mobilization of water management interested groups, then where are young people in this? For connecting 'disconnected' young adults growing up risk averse (see Gill 2007) in a climate change context will require radical forms of media exchange, civic engagement and social learning – well beyond *business as usual* and mainstream media communication. This kind of disruptive learning and engagement with young people through grassroots media may be commonplace in some Brazilian cities out of necessity but it is less evident in the UK context.

---

[13] Sendai Framework for Disaster Risk Reduction SFDRR 2015–2030 at http://www.data4sdgs.org.
[14] See Mayer-Schönberger and Cukier (2013), *Big Data: A Revolution That Will Transform How We Live, Work and Think.*
[15] See Jon Coaffee et al. (2008) for key research on urban resilience. All of which in the context of post-Covid recovery takes on a more intense need to listen to the less-heard voices in society.
[16] See Abers and Keck 2013 and Engle and Lemos 2010 on Brazil's unique and sometimes very effective river basin committee structure of participatory decision-making.

From an arts and humanities perspective, it has become clear to me that research on floods is dominated (rather obviously) by a focus on natural processes or designing hard-engineering defences, and not on transnational opportunities to exchange stories across research domains, participatory cultures and media forms. Moreover, most approaches to gathering flood data operate in a conventional way, as shown in the Hiscox advertisement at the beginning of this chapter: data generated by scientific and digital devices, sensors and citizen science, all sent to centralized systems for use by decision-makers, experts, scientists, governmental agents and other water authorities. Yet what of that other digitized space of data as story and memory? Here the active role of citizens, flood followers, water users, social movements, NGOs, participatory arts and media is engaged in generating, mapping and collecting stories and memories as a different kind of lived data, which could be more inclusive, multi-vocal and multi-directional (see Dodge and Kitchin 2013; Alevizou et al. 2016). A mapping of community assets which include the intangible cultural heritage of water memories and stories. How can we integrate these forms of flood data into scenarios of responsibility and collaboration?

Scenario-ing floods: Memories in(to) the future

In working with flood memories, remembrance, memorialization and mediatization it is clear that affective and emotional forms of knowledge are in operation. Recognizing this emotional value is the first step to understanding resilience. We know that it is through memory that persons and communities perform their senses of identity, their everyday lives, their watery or dry sense of place (alongside politicized versions of identity – class, race, ethnicity, religion, gender and sexuality, for example). Furthermore, in 'Materiality and Memory' Rigney has argued that objects (their shape, texture, colour and size) are 'accidental archives', that is, 'their materiality often secretes more meaning that that which was consciously inscribed in them' triggering 'searches and storytelling' and they 'demand to be looked after' (2017: 474). Thus, flood's materiality in place holds many of these memories of identity in a network, as accidental archives which cannot always be *data-ized*. A place of frequent flooding is both a storehouse of stories and images but also a space of transmission to its inhabitants who store those flood memories within themselves in and beyond that place, while maybe never sharing them. If we are to scenario flood risk and resilience then we need to rethink social relations to flooding, and possibly return to formations of past *flood-place-memory-story* relationships but through smarter knowledge networking. Communities may receive targeted (and even personalized) knowledge on the likelihood of local flood events happening; warning signs; preparedness information; knowledge on the speed, form and duration of local flood; the degree of risk to life or domestic and business properties; and actions to take during a flood; expectations in the immediate and post-flood recovery phases, and even mental health support services during recovery phase. Yet if flood memory as emotional experience is not visible, ready to activate, share and sustain communities, all these rational knowledge frameworks will not be as effective.

For 'given memory's fundamental role in individual, family and other small collectives, and the always-present aspects of space, landscape and place within the memories of such (and vice versa), there is a sense that these "smaller scale" dynamics of geography and memory remain under-represented and less-considered within both memory studies, geography and other disciplines' (Jones and Garde-Hansen 2012: 3).

Enabling transformation through media, art and memory will need to address the role that scenarios and scenario-ing take in risk and adaptation to flood, landslide and general water risks. Drawing on the fact that the concept of 'scenario' in the arts and humanities originates in theatre as detailed in Chapter 7, wherein a scenario is a visual/textual plan of the scene (pinned to the back of the theatrical curtain) that the actors are about to perform on stage, scenario-ing takes on new creative potential. Biggs et al. (2007: np) note that:

> Scenario exercises that aim to promote dialog between stakeholders at different scales are particularly challenging. The logistics are complex, and participatory scenario exercises consume significant resources, with costs rising as the network of stakeholders involved broadens. The benefits are an increased appreciation of perspectives from other scales, and a greater consideration of cross-scale processes and trade-offs between scales. To achieve such cross-scale participation, one needs to consider how scenarios at a particular scale can include stakeholders from other scales, either as observers or participants within the exercises themselves to help shape the focal questions or as an audience for the outputs.

When centres of calculation, data and expertise[17] (scientific, governmental, administrative and increasingly incorporating citizen science) model scenarios for water risk, it is the multi- and inter-scalarity of risk that is key to their communication and participation. Yet it is so difficult to achieve this in practice.

One suggestion that follows, based on transformation workshop-planning on flood risk in Brazil, it is an integrative approach of hydrological modelling and the arts and humanities practices of storytelling methods. Scales ranging from 10 per cent, 20 per cent, 30 per cent, 40 per cent of, for example, increased water levels or drought indicators (where 10 per cent creates a risk-story that produces mild stress in the domains while 40 per cent may express a narrative of maximal stress), entail a need for risk to be communicated in ways that offer multi-scale and multi-actor scenario-ing. In working with communities at water risk (inundation, scarcity, pollution, for example) it makes sense to address this scenario-ing in three key ways. Firstly, *plausibility* (or what we could term *scenarios or storytelling for exploratory purposes*). Starting from the present or recent past, centres of expertise could create a (good or bad) scenario in the area of their particular knowledge (hydrology, education, early warning systems, geological surveying, emergency services, health, business, social services, etc.). This would include responsibility or knowledge (that can be easily verified) at three or four levels of change that cover increases/decreases in rainfall (which are linked

---

[17] Such as the British Geological Survey in the UK or CEMADEN in Brazil.

to the flood or drought-risk modelling from data-led centres of expertise). Cultural and media artists and creatives could, then, work with water-risk communities to creatively produce three or four *plausible stories* (scenarios) from their perspective and their location drawing on their knowledge, the modelled data and evidence available to them from the organizations and experts. They could then share these with other locations, groups and sectors.

Secondly, *probability* (or what we could term *scenarios or storytelling for supporting decisions*). Drawing on the 'worst-case scenario' and the 'best-case scenario' storylines the communities have created in stage one, creatives could with their community groups, create one future scenario (an imagined story that cannot be verified) but is offering a probable result. Thirdly, it is time to take the probable scenarios for decision-making and upscale or downscale these. In upscaling – stakeholders take the probable scenarios from stage two and map against broader factors (social, national, global, cultural change, for example). In downscaling – stakeholders could take their probable scenario from stage two and map it against specific/local factors (personal, urban/rural, businesses, neighbourhoods, family, schools, etc.). The creatives and cultural organizations are important because they can help assess which of these scenarios (plausible or probable) in this location with this community would be best to communicate to the public and how; using what kinds of communication channels, genres, images, texts and evidence.

If this process draws on memories, which it will if it involves imaginative perspectives on living, being, experiencing and sensing water, then this has a correlation with Kok et al. (2011) on scenarios. Here local scenarios using different methods of forecasting and backcasting exercises establish a story of the present situation (now), a story of the future (then), a story of the near future (soon) extrapolating current trends and a story of the near future backcast from a desirable endpoint (what will have been). Such inventive storytelling created in groups means that flood memories for and in the future can be selected from stakeholder narratives and benchmarked against modelled data and climate scenarios, even if the research to support such storyline and timeline methods never really mentions the role that personal, collective or cultural memories play in the process.[18] How then to ensure that media and cultural studies methods are waterproofed for future research collaborations?

In 'Coproducing flood risk knowledge: redistributing expertise in critical "participatory modelling"' Landström et al. (2011: 1617) suggest that 'competency groups' in which the aim is 'to harness the energy generated in public controversy and enable other than scientific expertise to contribute to environmental knowledge' can address the need for data to adapt to culture. I have used competency groups in my own research and they do create a space for hydro-citizenship to emerge, suggestive of a new social direction in flood-risk preparedness as found in Nye et al. (2011). Likewise,

---

[18] A deeper connection between science and art, media and water was afforded by Nettley et al. (2014) whose shared landscape visualizations from computer models of past and future flooding were shared with residents and stakeholders to build feedback into iterative development of a mixed-media film about climate change and flooding on the River Tamar in Cornwall, UK.

the same research group in Lane et al. (2011) seeks to 'do flood risk differently' by looking at different types of knowledge, deconstructing the 'suite of practices' that form the basis on which computer modelling (of floods) is taken as objective, reliable and credible. They offer a hybridized approach of the possible range of imagined projections created by a simulation that can become constrained by policy goals at a range of scales and norms within the scientific community. Such constraining of imagination and creativity are at the very heart of the arts and humanities research agenda. The capacity of any model, metadata, framework or protocol to support a variety of future flood responses is limited if local residents who have lived through previous flooding and have alternative solutions are not included in the modelling and mapping. Therefore, the digital use of media, technology and communication can offer the following, and some of these principles were covered in part in McEwen et al. (2016):

- Rethink the ways that flood-risk groups can build community capital about how flood memories and associated lay knowledge are materialized, assimilated, embedded and protected
- Capture multi-stakeholder awareness and build capacity in engaging with flood memory in communities
- Highlight community capital in elders who have flood memories and knowledge of coping with past floods
- Reconceptualize who might be included as key actors in local flood-risk management to include mass communications media, social media and NGOs formal/informal archives, as well as accidental archives of flood memory
- Rethink the potential of the Internet of Things (and persons) to connect people to a watery sense of place more intimately in informal and everyday ways

Increased resilience of floodplain groups through the intersection of vertical and horizontal communications about flood memory and lay flood knowledge will only work if those memories are captured, curated and recollected. Through the development of media infrastructures, genres, discourses and practices and through human-data interfaces that respect both human and water relations as deeply cultural, social and emotional, flood and drought memory can be waterproofed.

Conclusion

Water platforms for new epistemologies

In *Contextualising Disaster: Catastrophes in Context* (2016) Button and Schuller focus on narration and globalization, and in a digital age of internet-enabled media and social media platforms it would make sense to do this. 'We need to view disasters' they say 'more as routine, normal, and connected to one another along various fault lines and as a direct product of culture and not something to be imagined as simply exceptional events'. For 'this allows us to move beyond the bounded unit of analysis that has too often typified traditional disaster studies and popular media narratives' (Button and Schuller 2016: 1). I would argue that a deeper and more diverse understanding of the history of the transnational and transcultural circulation of media narratives, media narration and the use of mediated storytelling towards environmental disruption would show that like water, media flows across many boundaries but are produced in specific contexts. There has been an intimacy between media and water for decades: such as maximal stresses fictionalised as overpowering water events in cinema. Social media platforms (such as Facebook, Instagram, Tencent and Weibo) are like the seas and reservoirs into which news, images and videos flow like rivers and streams. The question is, who controls the seas and reservoirs and where are all those water stories being stored?

It is clear that water will become an increasingly politicized and a culturally diverse and divisive issue in the mediated public sphere within and across many nations in the twenty-first century. In the case of recent catastrophic flood events in the last decade, the control of what we will term the elite production and consumption of flood knowledge by governments, agencies and national broadcast media has seen the emergence of a more visible, distributed water awareness through social networks drawing upon mobile and situated memories (such as Flickr, Facebook, YouTube, Twitter, etc.). This suggests that there are new communication technologies for remembering and connecting water. They are emerging not simply as an extension of broadcast infrastructure but emotionally and personally through sustaining connections with a watery sense of place and identity.

Globally we are at a 'watershed' in urban river research of 'water quality' (physical, biological, chemical) wherein river health reflects (globally connected) land use and past and present legacies and interventions (Meybeck 2005). Recent developments in memory studies (as in social psychology, digital studies, natural history and ecology) suggest moving the focus to an exploration of memory embedded and shaped in the social and natural environment that is guided and mediated by available resources. Memory studies research needs to be mobilized further to incorporate into its epistemology the cultural value of water (see Westling et al. 2014) in ways that are not simply addressing the materialization of human histories but the imagined futures of water. 'Productive mutual learning' from the Global South reports UNESCO (2016) could help address the personal, local, national, global routines and habits that *keep us locked in to neglectful practices*. Studies in Brazil have shown that, traditionally, urban waters have been engineered and canalized to hide problems related to the disposal of sewage, industrial pollution and urban development goals (Sakai and Frota 2014; Horne et al. 2018). Thus, if urban rivers reflect both existential terrains of mediated memory (Lagerkvist 2017) and cultural and historical uses (Zeisler-Vralsted 2015), we need to explore the 'power and the limits of stories' (Solinger et al. 2008) to understand urban riparian nurture-neglect, and be culturally ready for extreme natural events.

Brian Massumi defines extreme weather as 'the suddenly irrupting, locally self-organizing, systemically self-amplifying threat of large-scale disruption' (2011: 20). Like masses of people protesting in the streets, or lockdowns, extreme weather has massive socioeconomic consequences. More broadly, Brace and Geoghegan (2011), writing from the perspective of human geography, argue that climate change is encountered holistically, not just in how it is understood 'top-down' through the communication of scientific discourses but relationally at a local level: 'Climate change can be observed in relation to landscape but also felt, sensed, apprehended emotionally as part of the fabric of everyday life in which acceptance, denial, resignation and action co-exist as personal and social responses to the local manifestations of a global problem' (Brace and Geoghegan 2011: 284). Today, young people are emotionally and technologically connected, and these connections can be mobilized through sharing water knowledge on new platforms and across continents. But, this assumes the next generation is interested in water and that they will want to mediate water issues in established ways.[1]

Is water in a state of oblivion? Is it being remembered only for its violence or scarcity and forgotten for its everyday life-enhancing storytelling? Can it only be appreciated as contained (by territories and concrete)? To create *liquid memories* of water is to speak back to the (violent) riverbed of hard facts (economics, capitalism, power politics, environmental risk, urban development) and offer soft processes that work through the (largely patriarchal) past to produce trans and mobile modalities of uncontained flows of river stories. Water in oblivion (as the connection between remembering and forgetting says Marc Augé) is the necessary work state to understand river life narratives as self-realizing personal agency in a cultural policy of water. Here '[m]emories are like plants: there are those that need to be quickly eliminated in order to help the others burgeon, transform, flower' (Augé 2004: 17). The liquid memories proposed at the end of this book, as imagined scenarios the river, the flood and the

---

[1] In 'Where's the Water? Published on YouTube 5 May 2015, the YouTuber Benjamin Ferguson presents an animation in which two American 'dudes' discuss water consumption (in the context of the Californian drought) with an intelligent cow who communicates detailed statistics. The cow's head is then brutally chopped off by a 'dumb farmer' to shut him up. The severed cow's head continues to communicate water scarcity. Science should be able to meet art in disruptive and disturbing ways. Available at: https://www.youtube.com/watch?v=APjh4rONU4A&app=desktop. Accessed 1 September 2017.

drought could offer the possibility of circulating stories to and from the margins, banking cultural values of water for future generations out of possible stories of the past as well as alternative futures. Quite simply, this means remembering the future and remembering materiality.[2] As the feminist Karen Barard has argued in her interview with Dolphijn and van der Tuin: 'the "past" was never simply there to begin with, and the "future" is not what will unfold, but "past" and "future" are iteratively reconfigured and enfolded through the world's ongoing intra-activity'. Moreover, if as I argue memories of water disrupt the present and water is remembered through its eruption and materialized as media and culture, then 'the past, like the future [...] is not closed' for 'the "past" is open to change'.

> It can be redeemed, productively reconfigured in an iterative unfolding of space-time-matter. But its sedimenting effects, its trace, cannot be erased. The memory of its materializing effects is written into the world. So changing the past is never without costs, or responsibility.
>
> (Dolphijn and van der Tuin 2013: 66–7)

If much of the academic literature available on 'water' is couched in and framed by disciplines such as hydrology, engineering, environmental science, ecology, chemistry, geography and business studies, then how can water have a future if its pasts are so limited in our understanding? As a researcher in media, communication and cultural studies, I believe that my discipline is moving rather slowly in deeply engaging with the sciences and social sciences and that its direction remains at the level of politics, representation and exploring levels of sovereignty (producer, consumer, state, platform and user). These are all important, of course, but only touch the surface of environmental issues and crises and their mediated eventhood.

More specifically, in the wider arts and humanities disciplines, where I position my own research, there is less attention to the pragmatic use of media and culture for storying water and discussing water's precariousness and power through reference to media representations. In memory studies (my other field of expertise) there is a similar lack of attention to non-contiguous disciplinary research, drawing the arts and sciences together into productive dialogic exchange with environmental science. Media studies itself tends to address much wider global issues such as 'climate change' or 'sustainability' with less empirical evidence from place-based research. During the Covid19 pandemic of 2020 many micro-narratives of water emerged and were shared through social and mainstream media (river walks, rediscoveries of water play in the garden, the global sharing of photos of empty beaches and tourist-free coast-lines and a keen interest in wetland wildlife as just a few examples of

---

[2] Barard argues in *Meeting the Universe Halfway* (2007: 66) that what is needed is 'a robust account of the materialization of all bodies – "human" and "non-Human" – including the agential contributions of all material forces (both "social" and "natural"). This will require an understanding of the nature of the relationship between discursive practices and material phenomena: an accounting of "non-human" as well as "human" forms of agency; and an understanding of the precise causal nature of productive practices that take account of the fullness of matter's implication in its ongoing historicity.'

members of the pubic reconnecting with their local environments). In the meantime, water stakeholders (from flood-risk communities to farmers, from environmental agencies to water companies) are practising and making decisions in and through these media forms (through online discussions, sharing videos and photos, engaging with news organizations) with little time to reflect upon the discourses they produce about water, and thus can be unaware that mediating water impacts decisions and influences or even ignores the public. Furthermore, the public has access to a much richer repertoire of media images, texts, sounds and stories of water from around the world, and this is not being connected to local water uses by those with the means of cultural production.

Thus, if we are to develop a transnational understanding of the personal, cultural, collective and mediated memories of living with and without water (flood, drought, storm, rivers, seas, weather) then we need to better understand the role and potential of life-story, narrative and mnemonic practices (through a variety of media and materials). How communities experience their watery environment in terms of good, bad, emerging, erupting, corrupting, polluted, social and playful water courses is part of the multimedia mix of stories for engagement in a 'water universe'. Water researchers seek to provide a distinctive, innovative and transnational research perspective on dry, watery, flooded and hidden river-scapes. To achieve this they need personal, affective and community experiences to be captured, represented and shared, in order to provide a collaborative platform for better understanding the role that media and storytelling can have in articulating the *cultural values of water and water as culturally valuable*. Common wisdom linked to community memory can be marshalled to increase awareness of water issues, and this awareness can be used to develop community resilience.

While a watery sense of place (performed through a wide variety of narratives) can transcend national borders and contingently connect nations, it seems that media only really come into focus when they communicate flood or drought risk when it rains or there is a heatwave or on an anniversary of the last deluge or drought. Powerful concepts of 'local', 'national', 'European', 'commonwealth', 'Global North' and 'Global South' are scaling water and culture in not always helpful ways while digitally mediated water memories can offer data of the past, present and future as multimodal. These memories can include crowd-sourced evidence, personal stories and accidental archival records as well as print news, factual documentary, broadcast drama, blogs, fictional films, television, and all this in a context of tweets, selfies, posts, hashtags, memes, GIFs, animations and citizen media. We will, then, need new and multidirectional kinds of water platforms for communicating water's pasts and futures.

References

Aaronovitch, Ben (2011) *Rivers of London*, London: Victor Gollancz

Abers, Naerea and Margaret Keck (2013) *Practical Authority: Agency and Institutional Change in Brazilian Water Politics*, Oxford: Oxford University Press

Abram, David (1996) *The Spell of the Sensuous. Perception and Language in a More-than-Human World*, New York: Vintage Books

Adger, Neil W., Hughes, Terry P., Folke, Carl., Carpenter, Stephen R. and Johan Rockstrom (2005) 'Social-Ecological Resilience to Coastal Disasters', *Science Magazine* 309, 1036–9

Ahearne, Jeremy (2009) 'Cultural Policy Implicit and Explicit: A Distinction and Some Uses' *International Journal of Cultural Policy* 15: 2, 141–53

Ahmed, S. (2008) 'Gender and Integrated Water Resources Management in South Asia' in K. Lahiri-Dutt (ed) *Water First*, New Delhi: Sage, pp. 185–201

Ahmed, S. (ed) (2005) *Flowing Upstream: Empowering Women through Water Management Initiatives in India*, Ahmedabad: Foundation India Books

Alevizou, Giota, Alexiou, Katerina and Theo Zamenopoulos (2016) *Making Sense of Assets: Community Asset Mapping and Related Approaches for Cultivating Capacities*, The Open University and AHRC

Allan, Stuart, Adam, Barbara and Cynthia Carter (eds) (2000) *Environmental Risks and the Media*, London: Routledge

Allan, Stuart (2002) *Media, Risk and Science*, Buckingham: Open University Press

Allen, Matthew (2014) *The Labour of Memory*, Basingstoke: Palgrave Macmillan

Alleyne, Brian (2014) *Narrative Networks: Storied Approaches in a Digital Age*, London: Sage

Allon, Fiona and Zoë Sofoulis (2006) 'Everyday Water: Cultures in Transition', *Australian Geographer* 37: 1, 45–55

Alston, Margaret and Jenny Kent (2008) 'The Big Dry: The Link between Rural Masculinities and Poor Health Outcomes for Farming Men', *Journal of Sociology* 44, 133–47

Alston, M. (1995) 'Women and Their Work on Australian Farms', *Rural Sociology* 60, 521–32

Amrith, Sunil (2020) *Unruly Waters: How Mountain Rivers and Monsoons Have Shaped South Asia's History*, London: Penguin

Andersen, Mark D. (2011) *Disaster Writing: The Cultural Politics of Catastrophe in Latin America*, Charlottesville, VI: University of Virginia Press

Anderson, Deb (2014) *Endurance: Australian Stories of Drought*, Collingwood, VIC: Csiro Publishing

Anderson, Deb (2010) 'Drought, Endurance and Climate Change "Pioneers": Lived Experience in the Production of Rural Environmental Knowledge', *Cultural Studies Review* 16: 1, 82–101

Anderson, Deb (2008) 'Drought, Endurance and "the Way Things Were": The Lived Experience of Climate and Climate Change in Mallee', *Australasian Humanities Review* 45, 67–81

Andrews, Molly, Squire, Corinne and Maria Tamboukou (2013) *Doing Narrative Research* (2nd Edition), London: Sage

Ang, Ien (2011) 'Unsettling the National: Heritage and Diaspora' in Helmut Anheier and Yudhishthir Raj Isar's (eds) *Heritage, Memory and Identity*, London: Sage, pp. 82–94

Anheier, Helmut K. and Yudhishthir Raj Isar (2011) *Cultures and Globalization Heritage, Memory and Identity* London: Sage

Annin, Peter (2018) *The Great Lakes Water Wars* Washington, DC: Island Press

Appadurai, Arjun (1996) *Modernity at Large: Cultural Dimensions of Globalization*, Minneapolis, MN: University of Minnesota Press

Appadurai, Arjun (1990) 'Disjuncture and Difference in the Global Cultural Economy', *Theory, Culture & Society* 7: 2, 295–310

Armitage, David, Bashford, Alison and Sujit Sivasundaram (eds) (2018) *Oceanic Histories*, Cambridge: Cambridge University Press

Arnold, Matthew (2004 [1869]) *Culture and Anarchy*, edited by Jane Garner, Oxford: Oxford World Classics

Arnold, T. C. (2009) 'The San Luis Valley and the Moral Economy of Water' in J. M. Whiteley, H. Ingram and R. W. Perry (eds) *Water, Place and Equity*, Harvard: MIT Press, pp. 37–6

Ashdown, Isabel (2013) *Summer of '76*, Brighton: Myriad Editions

Ashton, Kevin (2009) 'That "Internet of Things" Thing', *RFiD Journal*

Askins, Kye (2009) 'Crossing Divides: Ethnicity and Rurality', *Journal of Rural Studies*, 25: 4, 365–75

Assmann, Aleida and Linda Shortt (eds) (2012) *Memory and Political Change*, Basingstoke: Palgrave Macmillan

Assmann, Aleida (2008) 'Canon and Archive' in A. Erll and A. Nunning (eds) *Cultural Memory Studies: An International and Interdisciplinary Handbook*, Berlin: De Gruyter, pp. 97–107

Assmann, Jan (2008) 'Communicative and Cultural Memory' in A. Erll and A. Nunning (eds) *Cultural Memory Studies: An International and Interdisciplinary Handbook*, Berlin: De Gruyter, pp. 109–18

Assmann, Jan (1995) 'Collective Memory and Cultural Identity' in Assmann Jan and John Czaplicka (eds) *New German Critique Cultural History/Cultural Studies* 65, 125–33

Atkinson, David (2008) 'The Heritage of Mundane Places' in Brian Graham and Peter Howard (eds) *The Ashgate Research Companion to Heritage and Identity*, Farnham: Ashgate, pp. 381–96

Ashley, Richard, Berry, Gersonius and Horton Bruce (2020) 'Managing Flooding: From a Problem to an Opportunity', *Philosophical Transactions of the Royal Society A: Mathematical, Physical and Engineering Sciences* 378 [online].

Augé, Marc (2004) *Oblivion*, translated by Marjolijn de Jager, Minneapolis, MI: University of Minnesota Press

Avritzer, Leonardo (2009) *Participatory Institutions in Democratic Brazil*. Washington: Wilson Press/Johns Hopkins University Press

Baake, Ken and Charlotte Kaempf (2015) 'No Longer "Bullying the Rhine;" Giving Narrative a Place in Flood Management', *Environmental Communication*, 5: 4, 428–46

Baake, Ken (2013) 'Commentary: The legacy of Charlie Flagg: Narratives of Drought and Overcoming the Monster in West Texas Water Policy Debates', *Texas Water Journal* 4: 1, 78–92

Bacallao-Pino, Lazaro M. (2016) 'Transmedia Events: Media Coverage of Global Transactional Repertoires of Collective Action' in B. Mitu and S. Poulakidakos (eds)

Media Events: A Critical Contemporary Approach, Basingstoke: Palgrave Macmillan, pp. 189–206

Bachelard, Gaston (1983) *Water and Dreams: An Essay on the Imagination of Matter*, translated by Edith R. Farrell, Ann Arbor, MI: Braun-Brumfield Inc.

Bachmair, S. Svensson, C., Hannaford, J., Barker, L. J. and K. Stahl (2016) 'A Quantitative Analysis to Objectively Appraise Drought Indicators and Model Drought Impacts', *Hydrology and Earth System Sciences* 20, 2589–609

Bakker, Karen (2001) 'Paying for Water: Water Charging and Equity in England and Wales', *Transactions of the Institute of British Geographers* 26: 2, 143–64

Balch, Oliver (2014) 'New Technology Uses Social Media to Keep Track of Water Levels' *The Guardian*, 17 July 2014. Available at: https://www.theguardian.com/sustainable-business/technology-social-media-water-levels-business. Accessed 16 December 2015

Ball, Philip (2000) *H20: A Biography of Water*, London: Phoenix

Barard, Karen (2007) *Meeting the Universe Halfway: Quantum Physics and the Entanglement of Matter and Meaning*, Durham, NC: Duke University Press

Barnett, Cynthia (2015) *Rain: A Natural and Cultural History*, New York: Crown Publishing

Barrett, Jeff and Robin Turner (2009) *Caught by the River: A Collection of Words on Water*, London: Cassell Illustrated

Barthes, Roland (1977) *Image/Music/Text*, Essays selected and translated by Stephen Heath, London: Fontana Press

BBC (2012) *Drought 2012: An Inside Out Special* (TV Programme)

BBC (2004) *Crisis Command — Could You Run the Country? 'Flood'* (TV Programme)

BBC1 (2018) *Spy in the Snow* (TV Programme)

BBC1 (2017–18) *Blue Planet II* (TV Series)

BBC1 (2011–12) *Frozen Planet* (TV Series)

BBC1 (2001) *The Blue Planet* (TV Series)

BBC2 (1979) *Life on Earth* (TV Series)

Bear, Christopher and Sally Eden (2011) 'Thinking Like a Fish? Engaging with Nonhuman Difference through Recreational Angling', *Environment and Planning D: Society and Space* 29: 2, 336–52

Beebeejaun, Y., Durose, C., Rees, J., Richardson, J. and L. Richardson (2013) 'Beyond Text: Exploring Ethos and Method in Co-Producing Research with Communities', *Community Development Journal*, 49: 1, 37–53

Bell, David and Kate Oakley (2015) *Cultural Policy*, London: Routledge

Bennett, Oliver (2014) *Cultures of Optimism: The Institutional Promotion of Hope*, Basingstoke: Palgrave Macmillan

Berryman, Colin, Burgin, Shelley and Tony Webb (2014) 'A Multifaceted, Cultural Approach to Community Engagement: Case Studies in Urban Water Management' *International Journal of Environmental Studies* 71: 3, 1–13

Bhattacharya, Shaoni (2003) 'European Heatwave Caused 35,000 Deaths' *New Scientist* 'Daily News' 19 October 2003

Biggs, Reinette, Raudsepp-Hearne, Ciara, Atkinson-Palombo, Carol, Bohensky, Erin, Boyd, Emily, Cundill, Georgina, Fox, Helen, Ingram, Scott, Kok, Kasper, Spehar, Stephanie, Tengö, Maria, Timmer, Dagmar and Moniker Zurek (2007) 'Linking Futures across Scales: A Dialog on Multiscale Scenarios' *Ecology and Society* 12: 1, 17

The Blob (1958, directed by Irvin S. Yeaworth)

Bohensky, Erin. and Ann Leitch (2013) 'Framing the Flood: A Media Analysis of Themes of Resilience in the 2011 Brisbane Flood' *Regional Environmental Change* 14: 2, 475–88

Bond, Lucy, Craps, Stef and Pieter Vermeulen (eds) (2016) *Memory Unbound: Tracing the Dynamics of Memory Studies*, New York: Berghahn Books

Bower, Joseph and Clayton M. Christensen (1995) 'Disruptive Technologies: Catching the Wave' *Harvard Business Review* 73: 1, 43–53

Boyd, Candice P. and Christian Edwardes (eds) (2019) *Non-Representational Theory and the Creative Arts*, Basingstoke: Palgrave Macmillan

Boykoff, Maxwell T. (2011) *Who Speaks for the Climate? Making Sense of Mass Media Reporting on Climate Change*, Cambridge: Cambridge University Press

Boym, Svetlana (2001) *The Future of Nostalgia*, New York: Basic Books

Bozzo, Sam (Dir.) (2008) *Blue Gold: World Water Wars*

Brace, Catherine and Hillary Geoghegan (2011) 'Human Geographies of Climate Change: Landscape, Temporality and Lay Knowledges' *Progress in Human Geography* 35: 3, 284–302

Brecht, Berthold (2007) 'On Violence' in John Willett (ed) *Poems 1913–1956*, London: Methuen

Briggs, Asa (1995) *The History of Broadcasting in the United Kingdom: Volume IV: Sound and Vision*, Oxford: Oxford University Press

Brockington, Dan (2008) 'Powerful Environmentalisms: Conservation, Celebrity and Capitalism' *Media, Culture and Society* 30: 4, 551–68

Brookey, Robert and Jonathan Gray (2017) '"Not Merely Para": Continuing Steps in Paratextual Research' *Critical Studies in Media Communication* 34: 2, 101–10

Brown, Katie Pride (2017) 'Water, Water Everywhere (or, Seeing Is Believing): The Visibility of Water Supply and the Public Will for Conservation' *Nature and Culture* 12: 3, 219–45

Bruns, Axel and Jean E. Burgess (2011) 'The Use of Twitter Hashtags in the Formation of Ad Hoc Publics' in *European Consortium for Political Research Conference*, 25–7 August, University of Iceland, Reykjavik

Bryant, Lia and Jodie George (2016) *Water and Rural Communities: Local Politics, Meaning and Place*, London: Routledge

Bryant, Lia and Bridget Garnham (2013) 'Beyond Discourses of Drought: The Micro-Politics of the Wine Industry and Farmer Distress' *Journal of Rural Studies* 32, 1–9

Bryant, Lia and Barbara Pini (2011) *Gender and Rurality*, New York: Routledge

Bryant, Lia (1999) 'The Detraditionalization of Occupational Identities in Farming in South Australia', *Sociologia Ruralis* 39, 236

Buechler, Stephanie and Anne-Marie Hanson (2015) *The Political Ecology of Women, Water and Global Environmental Change*, London: Routledge

Building Futures (2007) Living with Water: Visions of a Flooded Future RIBA, London. Available at: http://www.buildingfutures.org.uk/assets/downloads/pdffile_57.pdf. Accessed 4 August 2016

Bulloch, John and Adel Darwish (1993) *Water Wars: Coming Conflicts in the Middle East*, London: Victor Gollancz

Burnham-Fink, Michael (2015) 'Creating Narrative Scenarios: Science Fiction Prototyping at *Emerge*' *Futures* 70, 48–55

Button, Geoffrey V. and Mark Schuller (eds) (2016) *Conceptualizing Disaster: Catastrophes in Context*, New York: Berghahn Books

Cabinet Office (2008) *The Pitt Review: Lessons Learned from the 2007 Floods*, Cabinet Office, London, UK

Canal and River Trust (2015) *Putting the Water into Waterways: Water Resources Strategy 2015–2020*

Capstick Stuart B. Demski, Christina Sposato, Robert G. Pidgeon, Nick F. Spence and Adam Corner (2015) *Public Perceptions of Climate Change in Britain Following the*

Winter 2013/2014 Flooding, the Understanding Risk Research Group, Report. Available at: http://orca.cf.ac.uk/74368/1/URG%2015-01%20Flood%20Climate%20report%20 1%20May%202015%20final.pdf. Accessed 12 November 2018

Carlin, Phyllis Scott and Linda M. Park-Fuller (2012) 'Disaster Narrative Emergent/cies: Performing Loss, Identity and Resistance' *Text and Performance Quarterly* 32: 1, 20–37

Castells, Manuel (2000) *The Rise of the Network Society: The Information Age*, Oxford: Blackwell

Caughie, John (2000) *Television Drama: Realism, Modernism and British Culture*, Oxford: Oxford University Press

Channel 4 (2014–18) *Britain's Wildest Weather* (TV Series)

Chapman, A. Daniel, Lickel, Brian and Ezra Markowitz (2017) 'Reassessing Emotion in Climate Change Communication' *Nature Climate Change* 7: 12, 850–2

Chapman, G., Kumar, K., Fraser, C. and I. Gaber (1997) *Environmentalism and the Mass Media: The North-South Divide*, London: Routledge

Chappells, Heather, Medd, William and Elizabeth Shove (2011) 'Disruption and Change: Drought and the Inconspicuous Dynamics of Garden Lives' *Social and Cultural Geography* 12: 7, 701–15

Chen, C., McLeod, J. and A. Neimanis (eds) (2013) *Thinking with Water*. Montreal, Canada: McGill, Queen's University Press

Ciancia, Mariana, Piredda, Francesca and Simona Venditti (2014) 'Shaping and Sharing Imagination: Designers and the Transformative Power of Stories' in *Interactive Narratives, New Media and Social Engagement International Conference*, by Moura Hudson, Ricardo Sternberg, Regina Cunha, Cecelia Queiroz, and Martin Zeilinger. Available from ResearchGate.net under a CC licence. Accessed 31 December 2018

Clandinin, D. Jean (ed) (2007) *Handbook of Narrative Enquiry: Mapping a Methodology*, London: Sage

Clark, Terry Nichols and Daniel Silver (2013) 'Chicago from the Political Machine to the Entertainment Machine' in Carl Grodach and Daniel Silver (eds) *The Politics of Urban Cultural Policy: Global Perspectives*, London: Routledge, pp. 28–41

Clark, Wilma, Couldry, Nick, MacDonald, Richard and Hilde Stephansen (2014) 'Digital Platforms and Narrative Exchange: Hidden Constraints, Emerging Agency' *New Media and Society* 17: 6, 919–38

Clarke, Gillian (2009) *A Recipe for Water*, Manchester: Carcanet Press Ltd

Coaffee, Jon, Wood, David Murakami and Peter Rogers (2008) *The Everyday Resilience of the City: How Cities Respond to Terrorism and Disaster*, Basingstoke: Palgrave Macmillan

Coates, Peter (2013) *A Story of Six Rivers: History, Culture and Ecology*, London: Reaktion Books

Cohen, Bernard (1963) *The Press, the Public, and Foreign Policy*, Princeton, NJ: Princeton University Press

Cohen, Margaret (2010) *The Novel and the Sea (Translation/Transnation)*, Princeton, NJ: Princeton University Press

Collins, Kevin, Blackmore, Chris, Morris, Dick and Drennan Watson (2007) 'A Systemic Approach to Managing Multiple Perspectives and Stakeholding in Water Catchments: Some Findings from Three UK Case Studies' *Environmental Science and Policy* 10, 564–74

Cook, Bernic (2015) *Flood of Images, Media, Memory and Hurricane Katrina*, Austin, TX: University of Texas Press

Connerton, Paul (2008) 'Seven Types of Forgetting' *Memory Studies* 1, 59–71

Connick, Sarah and Judith E. Innes (2003) 'Outcomes of Collaborative Water Policy Making: Applying Complexity Thinking to Evaluation' *Journal of Environmental Planning and Management* 46: 2, 177–97

Constant, Natasha and Liz Roberts (2017) 'Narratives as a Mode of Research Evaluation in Citizen Science: Understanding Broader Science Communication Impacts' *Journal of Science Communication* 16: 4, 1–18

Corner, J., Richardson, K. and N. Fenten (1990) *Nuclear Reactions: Form and Response in 'Public Issue' Television*, London: John Libbey

Cotterill, Joseph (2018) 'South Africa: How Cape Town Beat the Drought' in *Huffington Post* 2 May 2018. Available at: https://highline.huffingtonpost.com/articles/en/cape-town-drought/. Accessed 1 June 2018

Couldry, Nick, Hepp, Andreas and Friedrich Krotz (eds) (2010) *Media Events in a Global Age* London: Routledge

Cox, Evelyn (1978) *The Great Drought of 1976*, London: Abe Books

Cox, Robert (2006) *Environmental Communication and the Public Sphere*, London: Sage

Craft, Aimée (2017) 'Giving and Receiving Life from Anishinaabe Nibi Inaakonigewin (our water law) Research' in J. Thorpe, S. Rutherford and A. Sandberg (eds) *Methodological Challenges in Nature-Culture and Environmental History Research*, London: Routledge, pp. 105–19

Craps, Stef, Crownshaw, Rick, Wenzel, Jennifer, Kennedy, Rosanne, Colebrook, Claire and Vin Nardizzi (2018) 'Memory Studies and the Anthropocene: A Roundtable' *Memory Studies* 11: 4, 498–515

Cresswell, Tim (2012) *Geographic Thought: A Critical Introduction*, London: John Wiley & Sons

Crow, Deserai and Michael Jones (2018) 'Narratives as Tools for Influencing Policy Change', *Policy and Politics* 46: 2, 217–34

Cruikshank, Joan (2005) *Do Glaciers Listen? Local Knowledge Colonial Encounters and Social Imagination*, Vancouver: UBC Press

Cubitt, Sean (2016) *Finite Media: Environmental Implications of Digital Technologies*, Durham, NC: Duke University Press

Cubitt, Sean (2009) *Ubiquitous Media, Rare Earths: The Environmental Footprint of Digital Media, Pervasive Media Lab*, Bristol: University of West of England

Cubitt, Sean (2005) *EcoMedia*, Amsterdam: Rodopi

Curran, James and Jean Seaton (2018) *Power without Responsibility: Press, Broadcasting and the Internet in Britain* (8th Edition), Abingdon: Routledge

Cusack, Tricia (2010) *Riverscapes and National Identities*, Syracuse: Syracuse University Press

Cusack, Tricia (2007) 'Introduction Riverscapes and the Formation of National Identity', *National Identities* 9: 2, 101–4

Cutter, S. L., Barnes, L., Berry, M., Burton, C., Evans, E., Tate, E. and J. Webb (2008) 'A Place-Based Model for Understanding Community Resilience to Natural Disasters' *Global Environmental Change* 18: 4, 598–606

da Cunha, Dilip (2019) *The Invention of Rivers: Alexander's Eye and Ganga's Descent*, Philadelphia: University of Pennsylvania Press

Dahlstrom, Michael F. (2014) 'Using Narratives and Storytelling to Communicate Science with Nonexpert Audiences' in *National Academy of Sciences of the United States of America*. Available at: https://www.pnas.org/content/pnas/early/2014/09/10/1320645111.full.pdf. Accessed 8 July 2018

Dalla Costa, Mariarosa and Monica Chilese (2014) *Our Mother Ocean: Enclosure, Commons, and the Global Fishermen's Movement*, translated by Silvia Federici, New York: Common Notions

Davies, Ann and Lorena Allam (2019) 'When the River Runs Dry: The Australian Towns Facing Heatwave and Drought', *The Guardian*, 25 January 2019

Davies, Caitlin (2015) *Downstream: A History and Celebration of Swimming the River Thames*, London: Arum Press

Dayan, Daniel and Elihu Katz (1992) *Media Events: The Live Broadcasting of History*, Cambridge, MA: Harvard University Press

Dayrell, Carmen and John Urry (2015) 'Mediating Climate Politics: The Surprising Case of Brazil' *European Journal of Social Theory* 18: 3, 257–83

De Cesari, Chiara and Ann Rigney (eds) (2014) *Transnational Memory: Circulation, Articulation, Scales*, Berlin: De Gruyter

De Landa, Manuel (2006) *A New Philosophy of Society: Assemblage Theory and Social Complexity*, London: Continuum

De Lisle, J., Goldstein, A. and G. Yang. (eds) (2016) *The Internet, Social Media, and a Changing China*, Philadelphia: University of Pennsylvania Press

De Silvey, Caitlin, Naylor , Simonand and Colin Sackett (2011) *Anticipatory History*, Uniform Books

De Sousa Santos, Boaventura (2014) *Epistemologies of the South: Justice against Epistemicide*, London: Routledge

Deakin, Roger (1999) *Waterlog: A Swimmer's Journey through Britain*, London: Chatto and Windus

Defoe, Daniel (2005 [1703]) *The Storm*, London: Penguin

Degrossi, Lívia Castro, De Albuquerque, Joao P., Fava, Maria Clara and Eduardo M. Mendiondo (2014) 'Flood Citizen Observatory: A Crowdsourcing-Based Approach for Flood Risk Management in Brazil' in 26th *International Conference on Software Engineering and Knowledge Engineering.*

Del Toro, Guillermo (Dir.) (2017) *The Shape of Water*

Deloughrey, Elizabeth (2017) 'Submarine Futures of the Anthropocene' *Comparative Literature* 69: 1, 32–44

Devitt, Catherine and Eoin O'Neill (2016) 'The Framing of Two Major Flood Episodes in the Irish Print News: Implications for Societal Adaptation to Living with Flood Risk' *Public Understanding of Science* 26: 7, 872–88

Dicks, Bella (2000) *Heritage, Place and Community*, Cardiff: University of Wales Press

Dodge, Martin and R. Rob Kitchin (2013) 'Crowdsourced Cartography: Mapping Experience and Knowledge' *Environment and Planning A: Economy and Space* 45: 1 19–36

Dolphijn, Rick and Iris van der Tuin (2013) 'Interview with Karen Barard' New Materialism: Interviews and Cartographies, Open Humanities Press. Available at: http://openhumanitiespress.org/books/download/Dolphijn-van-der-Tuin_2013_New-Materialism.pdf. Accessed 7 November 2016

Doyle, J. (2007) 'Picturing the Clima(c)tic: Greenpeace and the Representational Politics of Climate Change Communication' *Science as Culture* 16, 129–50

Edwards, Gareth A.S. (2013) 'Shifting Constructions of Scarcity and the Neoliberalization of Australian Water Governance' *Environmental Planning A* 45: 8, 1873–90

Edy, Jill A. (1999) 'Journalistic Uses of Collective Memory' *Journal of Communication* 4: 2, 71–85

Emmerich, Roland (Dir.) (2004) *The Day after Tomorrow*

Endfield, Georgina H. and Lucy Veale (eds) (2018) *Cultural Histories, Memories and Extreme Weather: A Historical Geography*, London: Routledge

Endtar-Wada, Joanna, Selfer, Theresa and Lisa W. Welsh (2009) 'Hydrologic Interdependencies and Human Cooperation: The Process of Adapting to Droughts' *Weather, Climate and Society* 1, 54–70

Engle, Nathan L. and Maria Carmen Lemos (2010) 'Unpacking Governance: Building Adaptive Capacity to Climate Change of River Basins in Brazil' *Global Environmental Change* 20: 1, 4–13

Escobar, Maria P. and David Demeritt (2014) 'Flooding and the Framing of Risk in British Broadsheets, 1985–2010' *Public Understanding of Science* 23: 4, 454–71

ESRC Sustainable Flood Memories project www.esrcfloodmemories.wordpress.com

European Union Water Framework Directive (2013) DIRECTIVE 2013/39/EU OF THE EUROPEAN PARLIAMENT AND OF THE COUNCIL of 12 August 2013

Evernden, Neil (1996) 'Beyond Ecology: Self, Place and the Pathetic Fallacy' in C. Glotfelty and H. Fromm (eds) *The Ecocriticism Reader: Landmarks in Literary Ecology*, Athens, GA: University of Georgia Press, pp. 92–104

Feldman, David Lewis (2012) *Water*, Cambridge: Polity Press

Fish, Robert, Church, Andrew, Michael Winter (2016) 'Conceptualising Cultural Ecosystem Services: A Novel Framework for Research and Critical Engagement' *Ecosystem Services* 21: B, 208–17

Fisher, Mark and Franco 'Bifo' Berardi (2013) 'Give Me Shelter', *Frieze*. Available at: https://frieze.com/article/give-me-shelter-mark-fisher. Accessed 10 July 2016

Florida, Richard (2004) *Cities and the Creative Class*, New York: Routledge

Folland, C. K., Hannaford, J., Bloomfield, J. P., Kendon, M., Svensson, C., Marchant, B. P., Prior, J. and E. Wallace Folland et al. (2015) 'Multi-Annual Droughts in the English Lowlands: A Review of Their Characteristics and Climate Drivers in the Winter Half-Year' *Hydrology and Earth System Sciences* 19, 2353–75

Forna, J., Fredriksson, M. and J. Johannisson (2009) 'What's the Use of Cultural Research' *Culture Unbound, Journal of Current Cultural Research*, 1, Linkoping University Electronic Press, 7–14

Foucault, Michel (2004/2008) *The Birth of Biopolitics: Lectures at the College de France 1978–1979*, translated by Graham Burchill, Basingstoke: Palgrave Macmillan

Furedi, Frank (2007) 'From the Narrative of the Blitz to the Rhetoric of Vulnerability' *Cultural Sociology* 1: 2, 235–54

Galaty, John (2010) 'How Visual Figures Speak: Narrative Inventions of "The Pastoralist" in East Africa' *Visual Anthropology* 15: 3, 347–67

Gandy, Matthew (2014) *The Fabric of Space: Water, Modernity and the Urban Imagination*, Harvard, MA: MIT Press

Garde-Hansen, Joanne and Gorton, Kristyn (2013) *Emotion Online: Theorising Affect on the Internet*, Basingstoke: Palgrave Macmillan

Garde-Hansen, Joanne, McEwen, Lindsey J., Holmes, Andrew and Owain Jones (2017) 'Sustainable Flood Memory: Remembering as Resilience' *Memory Studies* 10: 4, 384–405

Garde-Hansen Joanne, McEwen, Lindsey J. and Owain Jones (2016) 'Towards a Memo-techno-ecology: Mediating Memories of Extreme Flooding in Resilient Communities' in Andrea Hajek, Christine Lohmeier and Christian Pentzold (eds) *Social Memory in a Mediated World: Remembering in Troubled Times*, Basingstoke: Palgrave Macmillan, pp. 55–73

Garde-Hansen, Joanne, Hoskins, Andrew and Anna Reading (eds) (2009) *Save as ... Digital Memories*, Basingstoke: Palgrave Macmillan

Garde-Hansen, Joanne (2011) *Media and Memory*, Edinburgh: Edinburgh University Press

Geiß, Stefan (2018) 'The Dynamics of Media Attention to Issues' in P. Vasterman (ed) *From Media Hype to Twitter Storm News Explosions and Their Impact on Issues, Crises, and Public Opinion*, Amsterdam: Amsterdam University Press, pp. 83–113

Geiß, Stefan (2010) 'The Shape of News Waves. How the Intensity of Coverage of News Events Develops over Time'. Paper for the conference of the Association for Public Opinion Research (WAPOR) in Chicago. Mainz: Johannes Gutenberg-Universität

Genette, Gerard (1987) *Seuils*, translated into English in 1997 as *Paratexts: Thresholds of Interpretation*, Cambridge: Cambridge University Press

Genette, Gerard (1980) *Narrative Discourse: An Essay in Method*, translated by Jane E. Lewin, Ithaca, NY: Cornell University Press

Geyer, R., Jambeck, J. R. and K. Lavender Law (2017) 'Production, Use, and Fate of All Plastics ever Made' *Sciences Advances* 3: 7, 1–5

Giasson, Thierry, Brin, Colette and Marie-Michèle Sauvageau (2010) 'Le Bon, la Brute et le Raciste. Analyse de la couverture médiatique de l'opinion publique pendant la 'crise' des accommodements raisonnables au Québec' *Revue Canadienne de science politique* 43: 2, 379–406

Gibbons, Michael (1994) *The New Production of Knowledge: The Dynamics of Science and Research in Contemporary Societies*, London: Sage

Giddens, Anthony (1984) *The Constitution of Society: Outline of the Theory of Structuration*, Berkeley, CA, University of California Press.

Gill, Tim (2007) *No Fear: Growing Up in a Risk Averse Society*, London: Calouste Gulbenkian Foundation

Gilligan, Paula (2015) '"Blowtorch Britain": Labor, Heat and Neo-Victorian Values in Contemporary UK Media' in Julia Leyda and Diane Negra (eds) *Extreme Weather and Global Media*, London: Routledge, pp. 100–26

Gilroy, Paul (1993) *The Black Atlantic: Modernity and Double Consciousness*, London: Verso

Glasauer, Herbert (1998) 'Social Limitations of Sustainable Water Consumption' in J. Breuste, H. Feldmann and O. Uhlmann (eds) *Urban Ecology*, Berlin: Springer, pp. 311–14

Goldsmith, Ben, Ward, Susan and Tom O'Regan (2010) *Local Hollywood: Global Film Production and the Gold Coast*, Queensland: University of Queensland Press

Gonzales, Mike and Marianella Yanes (2015) *The Last Drop: The Politics of Water*, London: Pluto Press

Gorton, Kristyn and Garde-Hansen, Joanne (2019) *Remembering British Television: Audience, Archive, Industry*, London: BFI Bloomsbury

Gould, Pete and Rodney White (1974) *Mental Maps*, London: Penguin

Grand, Ann, Bultitude, Karen, Wilkinson, Clare and Alan F. T. Winfield (2010) 'Muddying the Waters or Clearing the Stream? Open Science as a Communication Medium' in *Public Communication of Science and Technology*, New Delhi, India, 6–9 December 2010. Available at: http://iscos.org/pcst/proceedings.htm. Accessed 18 June 2017

Graham-Leigh, John (2000) *London's Water Wars: The Competition for London's Water Supply in the Nineteenth Century*, London: Francis Boutle Publishers

Griffiths, H. M. and E. Salisbury (2013) '"The Tears I Shed Were Noah's Flood": Medieval Genre, Floods and the Fluvial Landscape in the Poetry of Guto'r Glyn' *Journal of Historical Geography* 40, 94–104

Griggs, Mary Beth (2018) '18 Water-Themed Books to Dive into This Month' Popular Science 2 August 2018. Available at: https://www.popsci.com/best-books-about-water. Accessed 18 November 2018

Grodach, Carl and Daniel Silver (eds) (2012) *The Politics of Urban Cultural Policy: Global Perspectives*, London: Routledge

Gross, C and David Dumaresq (2014) 'Taking the Longer View: Timescales, Fairness and a Forgotten Story of Irrigation in Australia' *Journal of Hydrology* 519, 2483–92

Groundwork (2016) *Jordan's Story* 'On the River'. YouTube. Available at: https://www.youtube.com/watch?v=f2FoFJ4vPOk

Gubbi Jayavardhana, Buyya, Rajkumar, Marusic, Slaven and Marimuthu Palaniswami (2013) 'Internet of Things (IoT): A Vision, Architectural Elements, and Future Directions' *Future Generation Computer Systems* 29, 1645–60

Hageman, Andrew (2009) 'Floating Consciousness: The Cinematic Confluence of Ecological Aesthetic in *Suzhou River*' in Sheldon H. Lu and Jiayan Mi (eds) *Chinese Cinema: In the Age of Environmental Challenge*, Hong Kong: Hong Kong University Press, pp. 73–93

Haigh and Bradshaw (2016) NERC's *Planet Earth* Series. Available at: https://nerc.ukri.org/planetearth/stories/1812/. Accessed 12 December 2018

Halbwachs, Maurice (1992) *On Collective Memory*, with and Coser, Lewis A, Chicago, MI: University of Chicago Press

Hall, Alexander (2018) 'Remembering in God's Name: The Role of the Church and Community Institutions in the Aftermath and Commemoration of Floods' in Georgina H. Endfield and Lucy Veale (eds) *Cultural Histories, Memories and Extreme Weather: A Historical Geography*, London: Routledge, pp. 112–32

Hall, Stuart (1994) 'Cultural Identity and Diaspora' in P. Williams and L. Chrisman (eds) *Colonial Discourse and Post-Colonial Theory: A Reader*, London: Harvester Wheatsheaf, 227–37

Hamlin, Christopher (1990) *A Science of Impurity: Water Analysis in Nineteenth Century Britain*, Berkeley, CA: University of California Press

Hansen, Anders (2010) *Environment, Media and Communication*, Abingdon: Routledge

Harvey, Colin (2014) 'A Taxonomy of Transmedia Storytelling' in M. Ryan and J. Thon (eds) *Storyworlds across Media: Toward a Media-Conscious Narratology*, Lincoln: University of Nebraska Press, pp. 278–94

Harvey, David and Jim Perry (eds) (2015) *Future of Heritage as Climates Change: Loss, Adaptation and Creativity*, Abingdon: Routledge

Hassan, John (1998) *A History of Water in Modern England and Wales*, Manchester: Manchester University Press

Hassid, Jonathan (2012) 'Safety Valve or Pressure Cooker? Blogs in Chinese Political Life' *Journal of Communication* 62: 2, 212–30

Hastrup, Frida (2008) 'Natures of Change: Weathering the World in Post Tsunami Tamil Nadu' *Nature and Culture* 3: 2, 135–50

Haughton, Graham, Bankoff, Greg and Tom J. Coulthard (2015) 'In Search of 'Lost' Knowledge and Outsourced Expertise in Flood Risk Management' *Transactions of the Institute of British Geographers* 40: 3, 375–86

Head, Dominic (2004) 'Ecocriticism and the Novel' in Laurence Coupe (ed) *The Green Studies Reader: From Romanticism to Ecocriticism*, London: Routledge, pp. 235–41

Head and Muir (2007) 'Changing Cultures of Water in Eastern Australian Backyard Gardens' *Cultural Geography* 8: 6, 889–905

Heise, Ursula K. (2002) 'Unnatural Ecologies: The Metaphor of the Environment in Media Theory' *Configurations* 10: 1, 149–68

Hervé-Bazin, Celine (2014) *Water Communication: Analysis of Strategies and Campaigns from the Water Sector*, London: IWA Publishing

Hewison, Robert (1998) The *Heritage Industry: Britain in a Climate of Decline*, London: Methuen

Hicks, Anna, Armijos, Maria Teresa, Barclay, Jenni, Stonea, Jonathan, Robertson, Richard and Gloria Patricia Cortése (2017) 'Risk Communication Films: Process, Product and Potential for Improving Preparedness and Behaviour Change' *International Journal of Disaster Risk Reduction* 23, 138–51

Hoare, Philip (2013) *The Sea Inside*, London: HarperCollins

Hochstetler, Kathryn and Margaret E. Keck (2007) *Greening Brazil: Environmental Activism in State and Society*, Durham, NC: Duke University Press

Holliman, Richard and Eric Jensen (2009) 'Investigating Science Communication to Inform Science Outreach and Public Engagement' in Richard Holliman, Elizabeth Whitelegg, Eileen Scanlon, Sam Smidt and Jeff Thomas (eds) *Investigating Science Communication in the Information Age: Implications for Public Engagement and Popular Media*, Oxford: Oxford University Press, pp. 55–71

Hood Washington, Sylvia, Rosier, Paul and Heather Goodall (eds) (2006) *Echoes from the Poisoned Well: Global Memories of Environmental Justice*, Lanham: Lexington Books

Horne, James, Tortajada, Cecilia and Larry Harrington (2018) 'Achieving the Sustainable Development Goals: Improving Water Services in Cities Affected by Extreme Weather Events' *International Journal of Water Resources Development* 34: 4, 475–89

Hoskins, Andrew (ed) (2018) *Digital Memory Studies: Media Pasts in Transition*, New York: Routledge

Hoskins, Andrew (2004) *Televising War from Vietnam to Iraq*, London: Continuum

Hossain, Amzad and Dora Marinova (2012) 'Grassroots Cultural Policy for Water Management in Bangladesh' *Water Practice and Technology* 7: 1. Available at: https://iwaponline.com/wpt

Howe, Cymene and Dominic Boyer (2020) 'Of Flood and Ice: Hydrological Globalization and the Rise of Water'. Keynote presented at *Flows and Floods: Changing Environments and Cultures Conference*, University of Warwick, UK, 22 February 2020

Hulme, Mike (2016) *Weathered: Cultures of Climate*, London: Sage

Hulme, Mike, Dessai, Suraje, Lorenzoni, Irene and Donald R. Nelson (2009) 'Unstable Climates: Exploring the Statistical and Social Constructions of "Normal" Climate' *Geoforum* 40, 197–206

Ingersoll, Karin Amimoto (2016) *Waves of Knowing: A Seascape Epistemology*, Durham, NC: Duke University Press

Ingold, Tim (2000) *The Perception of the Environment: Essays on Livelihood, Dwelling and Skill*, London: Routledge

Ioris, Antonio A. R. (2013) 'The Value of Water Values: Departing from Geography Towards an Interdisciplinary Synthesis' *Geografiska Annaler: Series B, Human Geography* 95: 4, 323–37

IPCC (2008) *Intergovernmental Panel on Climate Change*. Available at: http://www.ipcc-data.org/ddc_definitions.html

ITV (2018) *Wild Weather UK: Winners and Losers* (TV Programme)

Jacobs, Ronald N. (1996) 'Producing the News, Producing the Crisis: Narrativity, Television and News Work', *Media, Culture & Society* 18: 3, 373–97

Jansen, Bernard J., Zhang, Mimi, Sobel, Kate and Abdur Chowdury (2009) 'Twitter Power: Tweets as Electronic Word of Mouth' *Journal of the American Society for Information Sciences* 60: 11, 2169–88

Jencson, Linda (2000) 'Disastrous Rites: Liminality and Communitas in a Flood Crisis' *Anthropology and Humanism* 26: 1, 46–58

Jenkins, Henry (2008) *Convergence Culture: Where Old and New Media Collide*, New York: New York University Press

Jenkins, Henry (2006) *Fans, Bloggers, and Gamers: Exploring Participatory Culture*, New York: New York University Press

Jenkins, Henry, Ford, Sam and Joshua Green (2013) *Spreadable Media: Creating Value and Meaning in a Networked Culture*, New York: New York University Press

Jensen, Eric (2015) 'Highlighting the Value of Impact Evaluation: Enhancing Informal Science Learning and Public Engagement Theory and Practice' *Journal of Science Communication* 14: 3, 1–14

Jha, Alok (2015) *The Water Book*, London: Headline Publishing

Jones, Owain with Read, S. and John Wylie (2012) 'Unsettled and Unsettling Landscapes: Exchanges by Jones Read and Wylie about Living with Rivers and Flooding Watery Landscapes in an Era of Climate Change' *Journal of Arts and Communities* 4.1/4.2, 76–91

Jones, Owain and Katherine Jones (2017) 'On Narrative, Affect and Threatened Ecologies of Tidal Landscapes' in J. Thorpe, S. Rutherford and A. Sandberg (eds) *Methodological Challenges in Nature-Culture and Environmental History Research* London: Routledge, pp. 147–65

Jones, Owain and Joanne Garde-Hansen (eds) (2012) *Geography and Memory: Explorations in Identity, Place and Becoming*, Basingstoke: Palgrave Macmillan

Jones, Owain (2018) 'Towards Hydrocitizenship: Introduction and Overview'. Available at: https://www.hydrocitizenship.com/. Accessed 12 December 2018.

Jones, Owain (2011) 'Geography, Memory and Non-Representational Geographies' *Geography Compass* 5: 12, 1–11

Jones, Owain (2009) 'After Nature: Entangled Worlds' in Noel Castree, David Demeritt, Diana Liverman and Bruce Rhoads (eds) *A Companion to Environmental Geography*, Oxford: Blackwell, pp. 294–312

Jones, Roy and John H. Selwood (2012) 'From "Shackies" to Silver Nomads: Coastal Recreation and Coastal Heritage in Western Australia' in Ian Robertson (ed) *Heritage from Below*, Farnham: Ashgate, pp. 125–46

Jung, Kiju, Shavitt, Sharon, Viswanathan, Madhu and Joseph M. Hilbe (2014) 'Female Hurricanes Are Deadlier than Male Hurricanes', *Proceedings of the National Academy of Sciences of the United States of America* 111: 24, 8782–7

Kääpä, Pietari and Tommy Gustafsson (eds) (2013) *Transnational Ecocinemas: Film Culture in an Era of Ecological Transformation*, Bristol: Intellect

Kääpä, Pietari (2014) *Ecology and Contemporary Nordic Cinemas from Nation-building to Ecocosmopolitanism*, London: Bloomsbury

Katz, Elihu and Paul Lazarsfeld (1955) *Personal Influence: The Part Played by People in the Flow of Mass Communications*, New York: The Free Press

Katz, Elihu and Tamar Liebes (2007) '"No More Peace!": How Disaster, Terror and War Have Upstaged Media Events' *International Journal of Communication* 1, 157–66

Keightley, Emily, Pickering, Michael and Nicola Allett (2012) 'The Self-Interview: A New Method in Social Science Research' *International Journal of Social Research Methodology* 15: 6, 507–21

Keightley, Emily and Michael Pickering (2012) *The Mnemonic Imagination: Remembering as Creative Practice*, Basingstoke: Palgrave Macmillan

Kelly, Matthew (2017) '1953 Storm Surge: How Britain's Worst Natural Disaster Kicked Off the Debate on Climate Change' *The Conversation* 16 January 2017. Available at: https://theconversation.com/1953-storm-surge-how-britains-worst-natural-disaster-kicked-off-the-debate-on-climate-change-71310. Accessed 31 December 2018

Kellner, Douglas (2010) 'Media and Media Events: Some Critical Reflections' in N. Couldry et al. (eds) *Media Events in a Global Age*, London; New York: Routledge, pp. 76–91

Kelton, Elmer (1973) *The Time It Never Rained*, New York: Doubleday

Kempe, Michael (2003) 'Noah's Flood: The Genesis Story and Natural Disasters in Early Modern Times' in Michael Kempe and Christian Rohr (eds) *Coping with the Unexpected-Natural Disasters and Their Perception, Environment and History*, 9: 2, 151–71

Kenter, J. O., Raymond, C. M., van Riper, C. J. et al. (2019) 'Loving the Mess: Navigating Diversity and Conflict in Social Values for Sustainability' *Sustainability Science* 14, 1439–61

Klein, Naomi (2016) Lecture 'Let Them Drown: The Violence of Othering in a Warming World' (London). Available at: https://www.lrb.co.uk/v38/n11/naomi-klein/let-them-drown

Klein, Naomi (2012) 'The Most Important Thing in the World' in Amy Schrager Lang, Daniel Lang/ Levitsky (eds) *Dreaming in Public: Building the Occupy Movement*, Oxford: New Internationalist Publications, pp. 43–6

Klaebe, Helen G. (2013) 'Facilitating Local Stories in Post-Disaster Regional Communities: Evaluation in Narrative-Driven Oral History Projects' *Oral History Journal of South Africa* 1, 125–42

Knapton, O. and G. Rundblad (2014) 'Public Health in the UK Media: Cognitive Discourse Analysis and Its Application to a Drinking Water Emergency' in Hart and Cap (eds) *Contemporary Critical Discourse Studies*, London: Bloomsbury, pp. 559–82

Kok, Kasper, van Vliet, Mathijs, Bärlund, Ilona, Dubel, Anna and Jan Sendzimir (2011) 'Combining Participative Backcasting and Exploratory Scenario Development: Experiences from the SCENES Project' *Technological Forecasting and Social Change* 78, 835–51

Krause, Don and Mark Smith (2014) 'Twitter as Mythmaker in Storytelling: The Emergence of Hero Status by the Boston Police Department in the Aftermath of the 2013 Marathon Bombing' *Journal of Social Media in Society* 3: 1, 8–27

Krause, Franz, Garde-Hansen, Joanne and Nicola Whyte (2012) 'Flood Memories Media, Narratives and Remembrance of Wet Landscapes in England' *Journal of Arts and Communities* 4: 1, 128–42

Krause, Franz and Veronica Strang (2016) 'Thinking Relationships through Water' *Society and Natural Resources* 29: 6, 633–8

Krause, Franz (2017) 'Towards an Amphibious Anthropology of Delta Life' *Human Ecology* 45: 3, 403–8

Kunelius, R., Eide, E., Tegelberg, M., Yagodin, D. (2016) *Media and Global Climate Knowledge: Journalism and the IPCC*, New York: Palgrave Macmillan

Lagerkvist, Amanda (2017) 'Existential Media: Toward a Theorization of Digital Thrownness' *New Media and Society* 19: 1, 96–110

Lambert, D., Martins, L. and M. Ogborn (2006) 'Currents, Visions and Voyages: Historical Geographies of the Sea' *Journal of Historical Geography* 32, 479–93

Lambert, J. (2002) *Digital Storytelling Cookbook* (2nd Edition), Berkeley: Digital Diner Press

Landström, Catharina., Whatmore, Sarah J., Lane, Stuart N., Odoni, Nicholas A., Ward, Neil and Susan Bradley (2011) 'Co-producing Flood Risk Knowledge: Redistributing Expertise in Critical 'Participatory Modelling' *Environment and Planning A* 43, 1617–33

Landsberg, Alison (2004) *Prosthetic Memory: The Transformation of American Remembrance in the Age of Mass Culture*, New York: Columbia University Press

Lane, Stuart, N., Odoni, Nicholas A., Landstrom, Catharina, Whatmore, Sarah J., Ward, Neil and Susan Bradley (2011) 'Doing Flood Risk Science Differently: An Experiment in Radical Scientific Method' *Transactions of the Institute of British Geographers* 36, 15–36

Latour, Bruno (1988) *Science in Action: How to Follow Scientists and Engineers through Society*, Harvard, MA: Harvard University Press

Lavau, Stephanie (2013) 'Going with the Flow: Sustainable Water Management as Ontological Cleaving' *Environment and Planning D: Society and Space* 31: 3, 416–33

Lee, Jong Youl and Chad Anderson (2013) 'Cultural Policy and the State of Urban Development in the Capital of South Korea' in Carl Grodach and Daniel Silver (eds) *The Politics of Urban Cultural Policy: Global Perspectives*, London: Routledge, pp. 69–80

Leeson, Loraine (2018a) 'Water Power: Creativity and the Unlocking of Community Knowledge' in K. Jones and L. Roberts (eds) *Water, Meaning and Creativity: Understanding Human-Water Relationships*, London and New York: Routledge, pp. 23–35

Leeson, Loraine (2018b) 'Our Land: Creative Approaches to the Redevelopment of London's Docklands' in *International Journal of Heritage Studies*, Taylor and Francis, New York and London 25: 4

Leeson, Loriane (2014) 'Groundswell on the Thames: 30 Years of Cultural Activism with London's Riverside Communities' in WEAD-Women Environmental Artists Directory [online]. Available at: http://weadartists.org/groundswell-thames. Accessed 16 May 2016

Lefebvre, Henri (1991) *The Production of Space*, translated Donald Nicholson-Smith, Oxford: Blackwell

Lejano, Raul, Ingram, Mrill and Helen Ingram (2013a) *The Power of Narrative in Environmental Networks*, Cambridge, MA: MIT Press

Lejano, Raul, Tavares-Reager, J. and F. Berkes (2013b) 'Climate and Narrative: Environmental Knowledge in Everyday Life' *Environmental Science and Policy* 31, 61–70

Lester, L. and S. Cottle (2009) 'Visualizing Climate Change: Television News and Ecological Citizenship' *International Journal of Communication* 3: 1, 920–36

Lewis, Justin and Toby Miller (eds) (2003) *Critical Cultural Policy Studies: A Reader*, Oxford: Blackwell Publishers

Leyda, Julia and Diane Negra (eds) (2015) *Extreme Weather and Global Media*, London: Routledge

Leyshon, Michael and Jacob Bull (2011) 'The Bricolage of the Here: Young People's Narratives of Identity in the Countryside' *Social and Cultural Geography* 12: 2, 159–80

Linton, Jaimie and Jessica Budds (2014) 'The Hydrosocial Cycle: Defining and Mobilizing a Relational-Dialectical Approach to Water' *Geoforum* 57: 1, 170–80

Linton, Jamie (2010) *What Is Water? The History of a Modern Abstraction*, Vancouver: UBC Press

Littlefield, Robert S. and Andrea M. Quennette (2007) 'Crisis Leadership and Hurricane Katrina: The Portrayal of Authority by the Media in Natural Disasters' *Journal of Applied Communication Research* 35: 1, 26–47

Livingstone, Sonia (2009) 'On the Mediation of Everything' ICA Presidential Address 2008 *Journal of Communication* 59: 1, 1–18

Lockie, S., Lawrence, G. and L. Cheshire (2006) 'Reconfiguring Rural Resource Governance: The Legacy of Neo-liberalism in Australia' in P. Cloke, T. Marsden and P. Mooney (eds) *The Handbook of Rural Studies*, London: Sage, pp. 29–43

Lowe, Thomas, Brown, Katrina, Dessai, Suraje, Doria, Miguel de Franca, Haynes, Kat and Katharine Vincent (2006) 'Does Tomorrow Ever Come? Disaster Narrative and Public Perceptions of Climate Change' *Public Understanding of Science* 15: 1, 435–57

Lowenthal, D. (1988) *The Heritage Crusade and the Spoils of History*, Cambridge: Cambridge University Press

Lowenthal, David (1985) *The Past Is a Foreign Country*. Cambridge University Press, Cambridge

Lury, Celia and Nina Wakeford (eds) (2012) *Inventive Methods: The Happening of the Social*, London: Routledge

MacFarlane, Robert (2019) *Underland: A Deep Time Journey*, London: Hamish Hamilton

MacIntyre, Alasdair (2007) *After Virtue: A Study in Moral Theory* (3rd Edition), Paris: University of Notre Dame Press

MacKinnon, Rebecca (2008) 'Flatter World and Thicker Walls? Blogs, Censorship and Civic Discourse in China' *Public Choice* 134: 1–2, 31–46

Mackintosh, Sophie (2018) *The Water Cure*, London: Penguin

Maclean, K. and Woodward, E. (2012) 'Photovoice Evaluated: An Appropriate Visual Methodology for Aboriginal Water Resource Research' *Geographical Research* 51: 1, 94–105

Mahmoud, Mohammed et al. (2009) 'A Formal Framework for Scenario Development in Support of Environmental Decision-Making' *Environmental Modelling & Software* 24, 798–808

Mannheim, Karl (1959) 'The Problem of Generations' in Karl Mannheim (ed) *Essays on Sociology of Knowledge*, with Paul Kecskemeti (2nd Edition), London: Routledge & Kegan Paul, pp. 276–320

Mantel, Hillary (2003) *Giving Up the Ghost a Memoir*, London: Fourth Estate

Manzo, Kate (2010) 'Imaging Vulnerability: The Iconography of Climate Change' *Area: Royal Geographical Society with IBG* 42: 1, 96–107

March, Hug, Therond, Oliver and Delphine Leenhardt (2012) 'Water Futures: Reviewing Water-Scenario Analyses through an Original Interpretative Framework' *Ecological Economics* 82, 126–37

Marres, Noortje (2017) *Digital Sociology: The Reinvention of Social Research*, Cambridge: Polity

Marx, Karl (2007) *Economic and Philosophic Manuscripts of 1844*, New York: Dover

Massumi, Brian (2011) 'National Enterprise Emergency: Steps toward an Ecology of Powers' in Patricia T. Clough and Craig Willse (eds) *Beyond Biopolitics: Essays on the Governance of Life and Death*. Durham, NC: Duke University Press, pp. 19–45

Mathieson, Charlotte (ed) (2016) *Sea Narratives: Cultural Responses to the Sea, 1600 to the Present*, Basingstoke: Palgrave Macmillan

Matthews, Rachel (2017) *The History of the Provincial Press in England*, London: Bloomsbury

Mauch, Felix (2012) 'The Hamburg Flood in Public Memory Culture' *Arcadia* 4. Available at: http://www.environmentandsociety.org/arcadia/hamburg-flood-public-memory-culture. Accessed 5 September 2015

Mauss, Marcel (1990) *The Gift: The Form and Reason for Exchange in Archaic Societies*, London: Routledge

Maxwell, Richard and Toby Miller (2012) *Greening the Media*, Oxford: Oxford University Press

Maxwell, Richard and Toby Miller (2008) 'Ecological Ethics and Media Technology' *International Journal of Communication* 2, 331–53

Mayer-Schönberger, Viktor and Kenneth Cukier (2013) *Big Data: A Revolution That Will Transform How We Live, Work and Think*, London: John Murray

Mayer-Schönberger, Viktor (2011) *Delete: The Virtue of Forgetting in the Digital Age*, Princeton, NJ: Princeton University Press

McCombs, Maxwell E. and Donald L. Shaw (1972) 'The Agenda-Setting Function of Mass Media' *The Public Opinion Quarterly* 36: 2, 176–87

McConkey, James (1996) *The Anatomy of Memory*, Oxford: Oxford University Press

McCormack, Derek P. (2017) 'Elemental Infrastructures for Atmospheric Media: On Stratospheric Variations, Value and the Commons' *Environment and Planning D: Society and Space* 35: 3, 418–37

McCumber, Andrew (2017) 'Building "Natural" Beauty: Drought and the Shifting Aesthetics of Nature in Santa Barbara, California' *Nature and Culture* 12: 3, 246–62

McEwen, L., Garde-Hansen, J., Holmes, A., Jones, O. and F. Karuse (2016) 'Sustainable Flood Memories, Lay Knowledges and the Development of Community Resilience to Future Flood Risk' *Transactions of the Institute of British Geographers* 42: 1, 14–28

McEwen Lindsey J., Jones Owain and Iain Robertson (2014) '"A Glorious Time?" Reflections on Flooding in the Somerset Levels' *The Geographical Journal* 180, 326–37

McEwen Lindsey J., Iain, Robertson and Mike Wilson (2012a) 'Editorial – Learning to Live with Water: Flood Histories, Environmental Change, Remembrance and Resilience' *Journal of Arts and Communities* 4, 3–9

McEwen, Lindsey J., Reeves, D., Brice, J., Meadley, F. K., Lewis, K. and N. Macdonald (2012b) 'Archiving Flood Memories of Changing Flood Risk: Interdisciplinary Explorations around Knowledge for Resilience' *Journal of Arts and Communities* 4, 46–75

McEwen, Lindsey J. and Owain Jones (2012) 'Building Local/Lay Flood Knowledges into Community Flood Resilience Planning after the July 2007 Floods, Gloucestershire, UK' *Hydrology Research*, Special Issue 43, 675–88

McEwen, Lindsey J. (2011) 'Approaches to Community Flood Science Engagement: the Lower River Severn Catchment, UK as Case-study' *International Journal of Science in Society* 2: 4, 159–79

McEwen, Lindsey J. (2007) *Guidelines for Good Practice: Community Engagement with Local Flood Histories and Flood Risk*. Cheltenham: Lower Severn Community Flood Information Network, University of Gloucestershire

McGuigan, Jim (2014) 'The Neoliberal Self' *Culture Unbound* 6, 223–40

McGuigan, Jim (2004) *Rethinking Cultural Policy*, Maidenhead: Open University Press

Meadows, Daniel (2003) 'Digital Storytelling: Research-Based Practice' *New Media Visual Communication* 2: 2, 189–93

Meadows, Donella H. (2008) *Thinking in Systems: A Primer*, edited by Diana Wright, Sustainability Institute, White River Junction, VT: Chelsea Green Publishing

Medd, William, Deeming, H., Walker, G., Whittle, R., Mort, M., Twigger- Ross, C., Walker, M., Watson, N. and E. Kashefi (2015) 'The Flood Recovery Gap: A Real-Time Study of Local Recovery Following the Floods of June 2007 in Hull, North East England' *Journal of Flood Risk Management* 8, 315–28

Mellor, Mary (2009) 'Ecofeminist Political Economy and the Politics of Money' in Ariel Sallah (ed) *Eco-Sufficiency and Global Justice: Women Write Political Ecology*, London: Pluto Press, pp. 251–67

Mentz, Steven (2009) 'Toward a Blue Cultural Studies: The Sea, Maritime Culture, and Early Modern English Literature' *Literature Compass* 6: 5, 997–1013

Meybeck, Michel (2005) 'Looking for Water Quality' *Hydrological Processes* 19, 331–8

Meyrowitz, Joshua (1993) 'Images of Media: Hidden Ferment – and Harmony – in the Field' *Journal of Communication* 43: 3, 59–66

Michael, Mike (2012) 'Anecdote' in Celia Lury and Nina Wakeford (eds) *Inventive Methods: The Happening of the Social*, London: Routledge, pp. 25–36

Milbrandt, Tara (2017) 'Caught on Camera, Posted Online: Mediated Moralities, Visual Politics and the Case of Urban "Drought-shaming"' *Visual Studies* 32: 1, 3–23

Miller, Toby (2016) 'Cybertarian Flexibility – When Prosumers Join the Cognitariat, All That Is Scholarship Melts into Air' in M. Curtin and K. Sanson (eds) *Precarious Creativity: Global Media, Local Labour*, Oakland: University of California Press, pp. 19–32

Milmo, Cahal (2012) '"Crippling Drought" Hits South and East of England' *The Independent*, 21 February 2012. Available at: https://www.independent.co.uk/news/uk/home-news/crippling-drought-hits-south-and-east-of-england-7237415.html. Accessed 1 May 2018

Mitchell, David and Sharon Snyder (eds) (1997) *The Body and Physical Difference: Discourses of Disability*, Minneapolis, MI: University of Michigan Press

Mitu, Bianca and Stamtis Poulakidakos (eds) (2016) *Media Events: A Critical Contemporary Approach*, Basingstoke: Palgrave Macmillan

Mock, Cary J. (2018) 'The Temporal Memory of Major Hurricanes' in Georgina H. Endfield and Lucy Veale (eds) *Cultural Histories, Memories and Extreme Weather: A Historical Geography*, London: Routledge, pp. 78–92

Moran, Joe (2013) *Armchair Nation: An Intimate History of Britain*, London: Profile Books

Morgan, Ruth (2018) 'On the Home Front: Australians and the 1914 Drought' in Georgina H. Endfield and Lucy Veale (eds) *Cultural Histories, Memories and Extreme Weather: A Historical Geography*, London: Routledge, pp. 34–54

Morgan-Fleming, B., Riegle, S. and W. Fryer (2007) 'Narrative Inquiry in Archival Work' in D. Jean Clandinin (ed) *Handbook of Narrative Enquiry: Mapping a Methodology*, London: Sage pp. 81–97

Morrison, Toni (2008) *What Moves at the Margin: Selected Non Fiction*, Jackson: University of Mississippi Press

Moulaert, Frank, Martinelli, Flavia, Swyngedouw, Eric and Sara Gonzalez (eds) (2010) *Can Neighbourhoods Save the City? Community Development and Social Innovation*, London: Routledge

Mulvey, Laura (1999) 'Visual Pleasure and Narrative Cinema' in Leo Braudy and Marshall Cohen (eds) *Film Theory and Criticism: Introductory Readings*, New York: Oxford University Press, pp. 833–44

Murthy, Dhiraj and Scott A. Longwell (2013) 'Twitter and Disasters: The Uses of Twitter during the 2010 Pakistan Floods' *Information, Communication and Society* 16: 6, 837–55

Muzaini, Hamzah (2015) 'On the Matter of Forgetting and "Memory Returns"' *Transactions of Institute of British Geographers* 40: 1, 102–12

Myerhoff, B. (1982) 'Life History among the Elderly: Performance, Visibility and Remembering' in Jay Ruby (ed) *A Crack in the Mirror: Reflective Perspectives in Anthropology*, Philadelphia, PA: University of Pennsylvania Press, pp. 20–35

Myers, T. A., Nisbet, M. C., Maibach, E. W. and A. A. Leiserowitz (2012) 'A Public Health Frame Arouses Hopeful Emotions about Climate Change' *Climatic Change* 113: 3–4, 1105–12

Napoli, Philip (2011) *Audience Evolution: New Technologies and the Transformation of Media Audiences*, New York: Columbia University Press

Natural Resources Wales (2016–17) *State of Natural Resources Report*

Negra, Diane (ed) (2010) *Old and New Media after Katrina*, Basingstoke: Palgrave

Negrete, Aquiles (2013) 'Constructing a Comic to Communicate Scientific Information about Sustainable Development and Natural Resources in Mexico' *Procedia – Social and Behavioral Sciences* 103, 200–9

Neimanis, Astrida (2017) *Bodies of Water: Posthuman Feminist Phenomenology*, London: Bloomsbury

Netflix (2019) *Our Planet* (Documentary Series)

Nettley, Amy, Desilvey, Caitlin, Anderson, Karen, Wetherelt, Andrew and Chris Caseldine (2014) 'Visualising Sea-Level Rise at a Coastal Heritage Site: Participatory Process and Creative Communication' *Landscape Research* 39: 6, 647–67

Nicholson-Cole, S. (2005) 'Representing Climate Change Futures: A Critique on the Use of Images for Visual Communication' *Computers, Environment and Urban Systems* 29(1): 255–73

Nietzsche, Friedrich (2011) 'On the Uses and Disadvantages of History for Life' [1874] in Jeffrey K. Olick, Vered Vinitzky-Seroussi and Daniel Levy (eds) (2011) *The Collective Memory Reader*, Oxford: Oxford University Press, pp. 73–9

Norbury, Katherine (2015) *The Fish Ladder: A Journey Upstream*, London: Bloomsbury

Norman, Peggy (2020) 'What Comes after the Coronavirus Storm?' *The Wall Street Journal*, 23 April 2020

Nowotny, Helga, Scott, Peter and Michael Gibbons (2001) *Re-thinking Science: Knowledge and the Public in an Age of Uncertainty*, Cambridge: Polity Press

Nye, M., Tapsell, Sue and Clare Twigger-Ross (2011) 'New Social Directions in UK Flood Risk Management: Moving towards Flood Risk Citizenship?' *Journal of Flood Risk Management* 4: 288–97

Odih, Pamela (2014) *Watersheds in Marxist Ecofeminism*, Newcastle: Cambridge Scholars Press

OECD – Organisation for Economic Cooperation and Development (2001) 'Environmental Outlook', Available at: https://www.oecd-ilibrary.org/environment/oecd-environmental-outlook_9789264188563-en. Accessed 2 December 2017

O'Farrell, Maggie (2013) *Instructions for a Heatwave*, London: Headline Publishing

Olick, Jeffrey K., Vered Vinitzky-Seroussi and Daniel Levy (eds) (2011) *The Collective Memory Reader*, Oxford: Oxford University Press

Oliver-Smith, Anthony (2002) 'Theorizing Disasters: Nature, Power and Culture' in S. M. Hoffman and A. Oliver-Smith (eds) *Catastrophe and Culture: The Anthropology of Disaster*, Santa Fe, NM: School of American Research Press, pp. 23–47

Oliver-Smith, Anthony and Susanna Hoffmann (eds) (1999) *The Angry Earth: Disaster in Anthropological Perspective*, New York: Routledge

Onesto, Chris (2015) 'California Drought'. Available at: https://www.kcet.org/shows/artbound/art-water. Accessed 16 December 2018

O'Neill, Saffron and Neil Smith (2014) 'Climate Change and Visual Imagery' *WIREs Climate Change* 5: 1, 73–87

O'Neill, Saffron and S. Nicholson-Cole (2009) '"Fear Won't Do It" – Promoting Positive Engagement with Climate Change through Visual and Iconic Representation' *Science Communication* 30: 3, 355–80

O'Neill, Saffron (2013) 'Image Matters: Climate Change Imagery in US, UK and Australian Newspapers' *Geoforum* 49: 1, 10–19

O'Sullivan, Tim (1998) 'Nostalgia, Revelation and Intimacy: Tendencies in the Flow of Modern Popular Television' in Christine Geraghty and David Lusted (eds) *Television Studies Book*, London: Arnold, pp. 198–209

Orlove, B. and S. C. Caton (2010) 'Water Sustainability: Anthropological Approaches and Prospects' *Annual Review of Anthropology* 39: 401–15

Pahl-Wostl, C. (2008) 'Participation in Building Environmental Scenarios' in J. Alcamo (ed) *Environmental Futures: The Practice of Environmental Scenarios*, Amsterdam: Elsevier, pp. 105–22

Pamuk, Orhan (2012) *The Innocence of Objects/The Museum of Innocence Istanbul*, New York: Abrams

Pandurang, Mala (2001) 'Cross Cultural Texts and Diasporic Identities: Review of: Bromley, Roger' *Narratives for a New Belonging: Diasporic Cultural Fictions*', *Jouver: A Journal of Postcolonial Studies* 6, 1–2

Panelli, Ruth (2002) 'Contradictory Identities and Political Choices: "Women in Agriculture" in Australia' in Brenda Yeoh, Peggy Teo and Swee Lan Huang (eds) *Gender Politics in the Asia-Pacific Region*, London: Routledge, pp. 136–55

Parikka, Jussi (2012) *What Is Media Archaeology?* Cambridge: Polity Press

Parks, Lisa and Nicole Starosielski (eds) (2015) *Signal Traffic: Critical Studies of Media Infrastructures*, Urbana: University of Illinois Press

Paschen, Jana-Axinja and Raymond Ison (2014) 'Narrative Research in Climate Change Adaptation – Exploring a Complementary Paradigm for Research and Governance' *Research Policy* 43, 1083–92

Pasotti, Eleanora (2013) 'Brecht in Bogotá: How Cultural Policy Transformed a Clientalist Political Culture' in Carl Grodach and Daniel Silver (eds) *The Politics of Urban Cultural Policy: Global Perspectives*, London: Routledge, pp. 42–53

Paton, Douglas and David Johnston (2006) *Disaster Resilience: An Integrated Approach*, Springfield, IL: Charles C Thomas Pub Ltd

Payne Tom, Katherine, Jones and Owain Jones (2015) 'The Hydrocitizenship Project Celebrates World Water Day 2015' in J. M. Condie and Cooper (eds) *Dialogues of Sustainable Urbanisation: Social Science Research and Transitions to Urban Contexts*, Penrith: University of Western Sydney, pp. 236–40

Pearce, Cathryn (2018) 'Extreme Weather and the Growth of Charity: Insights from the Shipwrecked Fishermen and Mariners' Royal Benevolent Society, 1839–1860' in Georgina H. Endfield and Lucy Veale (eds) *Cultural Histories, Memories and Extreme Weather: A Historical Geography*, London: Routledge, pp. 55–77

Picken, Felicity and Tristan Ferguson (2014) 'Diving with Donna Haraway and the Promise of a Blue Planet' *Environment and Planning D: Society and Space* 32: 2, 329–41

Pickering, Michael and Emily Keightley (2013) 'Communities of Memory and the Problem of Transmission' *European Journal of Cultural Studies* 16: 1, 115–31

Pini, Maria (2001) 'Video Diaries: Questions of Authenticity and Fabrication' in Screening the Past. Available at: http://www.screeningthepast.com/2014/12/video-diaries-questions-of-authenticity-and-fabrication/. Accessed 12 January 2019

Pink, Sarah (2007) 'Visual Methods' in C. Seale, G. Gobo, J..F. Gubrium and D. Silverman (eds) *Qualitative Research Practice*, London, Sage, pp. 361–77

Pink, Sarah (2001) *Doing Visual Ethnography: Images, Media and Representation in Research*, London: Sage

Price, Linda and Nick Evans (2009) 'From Stress to Distress: Conceptualizing the British Family Farming Patriarchal Way of Life', *Journal of Rural Studies* 25, 1–21

Price, Linda and Nick Evans (2005) 'Work and Worry: Revealing Farm Women's Way of Life' in J. Little and C. Morris (eds) *Critical Studies of Rural Gender Issues*, Aldershot: Ashgate, pp. 45–59

Princess Mononoke (1997, directed Hayao Miyazaki)

Probyn, Elspeth (2016) *Eating the Ocean*, Durham, NC: Duke University Press

Quigley, Cassie and Gayle Buck (2012) 'The Potential of Photo-Talks to Reveal the Development of Scientific Discourses' *Creative Education* 3: 2, 208–16

Radstone, Susannah and Katharine Hodgkin (eds) (2005) *Memory Cultures: Memory, Subjectivity and Recognition*, Piscataway, NJ: Transaction Books

Ramirez-Ferrero, E. (2005) *Troubled Fields: Men, Emotions and the Crisis in American Farming*, New York: Columbia University Press

Reading, Anna and Tanya Notley (2015) 'The Materiality of Globital Memory: Bringing the Cloud to Earth' *Continuum: Journal of Media and Cultural Studies* 29: 4, 511–21

Reading, Anna and Ned Rossiter (2012) 'Data, Memory, Territory' *Digital Media Research* 1, 5

Reading, Anna (2014) 'Seeing Red: A Political Economy of Digital Memory' *Media, Culture and Society* 36: 6, 748–60

Reading, Anna (2012) 'The Dynamics of Zero: On Digital Memories of Mars and the Human Fœtus in the Globital Memory Field' *Journal for Communication Studies* 5: 2, 21–44

Reading, Anna (2010) 'Gender and the Right to Memory' *Media and Development* LVII: 2, 11–14

Rice, Alan (2012) 'The History of the Transatlantic Slave Trade: Heritage from below in Action: Guerrilla Memorialisation in the Era of Bicentennial Commemoration in Ian Robertson (ed) *Heritage from Below*, Farnham: Ashgate, pp. 209–36

Riessman, Catherine Kohler (2007) *Narrative Methods for the Human Sciences*, London: Sage

Rigney, Ann (2017) 'Materiality and Memory: Objects to Ecologies. A Response to Maria Zirra' *Parallax* 23: 4, 474–8

Rimmon-Kenan, Shlomith (2006) 'Concepts of Narrative' in Matti Hyvärinen, Anu Korhonen and Juri Mykkänen (eds) *The Travelling Concept of Narrative Studies across Disciplines in the Humanities and Social Sciences* 1, Helsinki: Helsinki Collegium for Advanced Studies, pp. 10–19

Robertson, Ian (ed) (2012) *Heritage from Below*, Farnham: Ashgate

Robertson, Ian (2008) 'Heritage from Below: Class, Social Protest and Resistance' in B. Graham and P. Howard (eds) *The Ashgate Research Companion to Heritage and Identity*, Farnham: Ashgate, pp. 143–58

Robinson, Sue (2009) '"We were All There": Remembering America in the Anniversary Coverage of Hurricane Katrina' *Memory Studies* 2: 2, 235–53

Rödder, Simone, Franzen, Martina, Peter Weingart (eds) (2012) *The Sciences' Media Connection – Public Communication and Its Repercussions*, Heidelberg: Springer

Roeser, Sabine (2012) 'Risk Communication, Public Engagement, and Climate Change: A Role for Emotions' *Risk Analysis: An International Journal* 32: 6, 1033–40

Rose, Gillian (2008) 'Using Photographs as Illustrations in Human Geography' *Journal of Geography in Higher Education* 32: 1, 151–60

Rothberg, Michael (2009) *Multidirectional Memory: Remembering the Holocaust in the Age of Decolonization*, Palo Alto, CA: Stanford University Press

Rundblad, Gabriella, Knapton, Olivia and Paul R. Hunter (2010) 'Communication, Perception and Behaviour during a Natural Disaster Involving a "Do Not Drink" and a Subsequent "Boil Water" Notice: A Postal Questionnaire Study' *BMC Public Health* 10, 641

Ruppert, Evelyn, Law, John and Mike Savage (2013) 'Reassembling Social Science Methods: The Challenge of Digital Devices' *Theory, Culture and Society* 30: 4, 22–46

Ryan, Marie-Laure and Jan-Noel Thon (eds) (2014) *Storyworlds across Media: Toward a Media-Conscious Narratology*, Lincoln: University of Nebraska Press

Said, Edward (1978) *Orientalism*, London: Pantheon Books

Sakai, Diogo Isao Santos and José Artur D'Aló Frota (2014) 'Águas Urbanas: Caminhos Para um Resgate/Urban Waters: Path to a Rescue' *III Seminário Nacional sobre o Tratamento de Áreas de Preservação Permanente em Meio Urbano e Restrições Ambientais ao Parcelamento do Solo será realizado em Belém do Para*, Belem: UFPA, 10–13 September 2014

Salina, Irena (Dir.) (2008) *Flow: For the Love of Water*

Salmon, Christian (2010) *Storytelling: Bewitching the Modern Mind*, London: Verso

Salvaggio, M., Futrell, R., Batson, C. and B. Brents (2014) 'Water Scarcity in the Desert Metropolis: How Environmental Values, Knowledge and Concern Affect Las Vegas Residents' Support for Water Conservation Policy' *Journal of Environmental Planning and Management* 57: 4, 588–611

Samuel, Raphael (1994) *Theatres of Memory*. Volume 1, London: Verso

Sandover, Rebecca (2014) 'While Ministers Dither on Floods, Social Media Springs into Action' *The Conversation*. Available at: https://theconversation.com/while-ministers-dither-on-floods-social-media-springs-into-action-23055. Accessed 13 October 2016

Sayers, Paul, Penning-Rowsell, Edmund C. and Matt Horritt (2018) 'Flood Vulnerability, Risk and Social Disadvantage: Current and Future Patterns in the UK' *Regional Environmental Change* 18: 2, 339–52

Schenk, Gerrit J. (2007) 'Historical Disaster Research: State of Research, Concepts, Methods and Case Studies' *Historical Social Research* 32: 3, 9–31

Schmidt, Jeremy and Kyle Mitchell (2013) 'Property and the Right to Water' *Review of Radical Political Economics* 46: 1, 54–69

Schuster, J. Mark (2002) *Informing Cultural Policy: The Research and Information Infrastructure*, New Brunswick, NJ: Center for Urban Policy Research, State University of New Jersey

Schwartz, J. (1996) 'The Geography Lesson: Photographs and the Construction of Imaginative Geographies' *Journal of Historical Geography* 22: 1, 16–45

Setten, Gunhild (2012) 'What's in a House? Heritage in the Making of the South-Western Coast of Norway' in Ian Robertson, (ed) *Heritage from Below*, Farnham: Ashgate, pp. 147–76

Setterfield, Diana (2018) *Once Upon a River*, London: Doubleday Books

Shanahan, James, Pelstring, Lisa and Katherine McComas (1999) 'Using Narratives to Think About Environmental Attitude and Behavior: An Exploratory Study', *Society & Natural Resources*, 12(5), 405–19

Sharp, Liz (2017) *Reconnecting People and Water: Public Engagement and Sustainable Urban Water Management*, Abingdon: Oxford University Press

Sharp, Liz, McDonald, Adrian, Sim, Patrick, Knamiller, Cathy, Sefton, Christin and Sam Wong (2011) 'Positivism, Post-positivism and Domestic Water Demand: Interrelating Science Across the Paradigmatic Divide', *Transactions of the Institute of British Geographers* 36: 4, 501–15

Sheppard, S. (2005) 'Landscape Visualisation and Climate Change: The Potential for Influencing Perceptions and Behaviour' *Environmental Science and Policy* 8: 1, 637–54

Sherren, K. and C. Verstraten (2012) 'What Can Photo-Elicitation Tell Us about How Maritime Farmers Perceive Wetlands as Climate Changes?' *Wetlands* 33: 1, 65–81

Shiva, Vandana (2016) *Water Wars: Privatization, Pollution and Profit*, Berkeley: North Atlantic Books

Shukman, David (2014) 'Barrage Over Climate Change Link to Floods' *BBC News On-line*. Available at: http://www.bbc.co.uk/news/science-environment-26242253. Accessed 9 January 2016

Sinclair, Iain (2004) *Downriver*, London: Penguin

Sklair, Leslie (2019) 'The Corporate Capture of Sustainable Development and Its Transformation into a "Good Anthropocene" Historical Bloc', *Civitas – Journal of Social Sciences*, 19(2), 296–314

Sklair, Leslie (2018) 'The Anthropocene Media Project: Mass Media on Human Impacts on the Earth System' *Visions for Sustainability* 10

Sklair, Leslie (2017) 'Sleepwalking through the Anthropocene (Review article)' *British Journal of Sociology* 68: 4, 775–84

Skoric, Marko M., Ping, Esther Chua Jia and Angeline Liew, Meiyan and Hui Wong, Keng and Pei Jue Yeo (2010) 'Online Shaming in the Asian Context: Community Empowerment or Civic Vigilantism?' *Surveillance and Society* 8: 2, 181–99

Sloterdijk, Peter (2011) *Bubbles: Spheres Volume 1*, Los Angeles, CA: Semiotext(e)

Smith, Benjamin, Clifford, Nicholas and Jenny Mant (2014) 'The Changing Nature of River Restoration' *WIREs Water* 1: 3, 249–61

Smith, Laura Jane (2006) *Uses of Heritage*, London: Routledge

Soechtig, Stephanie and Jason Lindsay (Dirs) (2009) *Tapped*

Solinger, Rockie, Madeline, Fox and Kayhan Irani (eds) (2008) *Telling Stories to Change the World: Global Voices on the Power of Narrative to Build Community and Make Social Justice Claims*, London: Routledge

Solomon, Steven (2010) *Water: The Epic Struggle for Wealth, Power and Civilization*, New York: HarperCollins

Somerville, Margaret (2013) *Water in a Dry Land: Place-Learning through Art and Story*, London: Routledge

Squire, Corinne (2008) *Approaches to Narrative Research*, ESRC National Centre for Research Methods Review Paper, National Centre for Research Methods NCRM Review Papers NCRM/009. Available at: http://eprints.ncrm.ac.uk/419/1/MethodsReviewPaperNCRM-009.pdf. Accessed 20 January 2019

Stedman, Lis (2014) 'Writing about Water: The Changing Face of Communication' in Celine Hervé-Bazin (ed) *Water Communication: Analysis of Strategies and Campaigns from the Water Sector*, London: IWA Publishing, pp. 125–6

Stehlik, Daniela, Lawrence, Geoffrey and Ian Gray (2000) 'Gender and Drought: Experiences of Australian Women in the Drought of the 1990s' *Disasters*, 24: 1, 38–53

Steinberg, Philip and K. Peters (2015) 'Wet Ontologies, Fluid Spaces: Giving Depth to Volume through Oceanic Thinking' *Environment and Planning D: Society and Space* 33, 247–64

Steinberg, Philip E. (2001) *The Social Construction of the Ocean*, Cambridge: Cambridge University Press

Stracher, Cameron (2011) *The Water Wars*, Naperville, IL: Sourcebooks Inc

Strang, Veronica (2014) 'Fluid Consistencies: Meaning and Materiality in Human Engagements with Water' *Archaeological Dialogues* 21: 2, 133–50

Strang, Veronica (2009) *Gardening the World: Agency, Identity and the Ownership of Water*, New York: Berghahn

Strang, Veronica (2005) 'Common Senses: Water, Sensory Experience and the Generation of Meaning' *Journal of Material Culture* 10: 1, 92–120

Strang, Veronica (2004) *The Meaning of Water*, Oxford: Berg

Strüver, Anke (2007) 'The Production of Geopolitical and Gendered Images through Global Aid Organisations' *Geopolitics* 12: 4, 680–703

Swift, Graham (1983) *Waterland*, London: Heinemann

Swyngedouw, Eric (2015) *Liquid Power Contested Hydro-Modernities in Twentieth-Century Spain*, Cambridge, MA: MIT Press

Swyngedouw, Eric (2009) 'The Political Economy and Political Ecology of the Hydro Social Cycle' *Journal of Contemporary Water Research and Education* 142: 1, 56–60

Swyngedouw Eric (1999) 'Modernity and Hybridity: Nature, Regeneracionismo, and the Production of the Spanish Waterscape, 1890–1930' *Annual Association of American Geographers* 89: 3, 443–65

Talbot, Bryan and Mary Talbot (2019) *Rain*, London: Vintage, Jonathan Cape

Talling, Paul (2011) *London's Lost Rivers*, London: Penguin

Tam, Yee-Lok (2012) 'Colourful Screens: Water Imaginaries in Documentaries from China and Taiwan' *Interactions: Studies in Communication and Culture* 2: 2, 109–26

Tapsell, Sue (1997) 'Rivers and River Restoration: A Child's Eye View' *Landscape Research* 22: 1, 45–65

Taylor, A., de Bruin, W. B. and S. Dessai (2014) 'Climate Change Beliefs and Perceptions of Weather-Related Changes in the United Kingdom' *Risk Analysis* 34: 11, 1995–2004

Taylor, Vanessa, Chappells, Heather, Medd, Will and Frank Trentmann (2009) 'Drought Is Normal: The Socio-Technical Evolution of Drought and Water Demand in England and Wales, 1893–2006' *Journal of Historical Geography* 35: 3, 568–91

Thelwall, Mike and Kevan Buckley (2013) 'Topic-based Sentiment Analysis for the Social Web: The Role of Mood and Issue-related Words', *Journal of the American Society of Information Science and Technology*, 64, 1608–17

Thompson, Paul (2017) *The Voice of the Past: Oral History* (4th Edition), with Joanna Bornat, New York: Oxford University Press

Thompson, William (2017) *Tides and the Ocean: Water's Movement around the World, from Waves to Whirlpools*, New York: Black Dog and Leventhal Publishers

Thrift, Nigel (2004) 'Intensities of Feeling: Towards a Spatial Politics of Affect' *Geografiska Annaler* 86 B: 1, 57–78

Thrift, Nigel (2000) 'Afterwords', *Environment and Planning D: Society and Space*, 18: 2, 213–55

Thrift, Nigel (1997) *Non-Representational Theory: Space, Politics, Affect*, London: Routledge

Thrift, Nigel (1992) 'Muddling through: World Orders and Globalization' *The Professional Geographer* 44: 1, 3–7

Thorpe, Jocelyn, Rutherford, Stephanie and L. Anders Sandberg (eds) (2017), *Methodological Challenges in Nature-Culture and Environmental History Research*, London: Routledge

Thussu, Daya Kishan (ed) (2007) *Media on the Move Global Flow and Contra-flow*, London: Routledge

Trümper, Stephanie and I. Neverla (2013) 'Sustainable Memory: How Journalism Keeps the Attention for Past Disasters Alive' *Studies in Communication Media* 2, 1–37

Tuohy, R. and C. Stephens (2012) 'Older Adults' Narratives about a Flood Disaster: Resilience, Coherence, and Personal Identity' *Journal of Aging Studies* 26: 1, 26–34

Tvedt, Terje (2015) *Water and Society: Changing Perceptions of Society and Historical Development*, London: I.B. Tauris

UK Environment Agency (2016) *Living on the Edge: A Guide to Your Rights and Responsibilities of Riverside Ownership*. Available at: https://www.gov.uk/guidance/owning-a-watercourse.

Ullberg, Susann Baez (2017) 'Forgetting Flooding?: Post-disaster Livelihood and Embedded Remembrance in Suburban Santa Fe, Argentina' *Nature and Culture* 12: 1, 26–45

Under the Skin (2013, directed by Jonathan Glazer)

United Nations (2015) *Sendai Framework for Disaster Risk Reduction 2015–2030*. Available at: https://www.undrr.org/publication/sendai-framework-disaster-risk-reduction-2015-2030

United Nations (2013a) *Water factsheets*. Available at: https://www.unwater.org/water-facts/

United Nations (2013b) *International Decade for Action 'Water for Life' 2005–2015*. Available at: https://www.un.org/waterforlifedecade/

United Nations (2012) *Passport to Mainstreaming Gender in Water Programmes*. Available at: http://www.fao.org/policy-support/tools-and-publications/resources-details/en/c/1260580/

UNESCO (2016) *Drought Risk Management: A Strategic Approach*, United Nations Educational, Scientific and Cultural Organization, Paris: UNESCO

UNESCO (2013) 'International Year of Water Co-operation'

UNESCO (2007) *Declaration on the Rights of Indigenous Peoples*. Available at: https://www.un.org/development/desa/indigenouspeoples/declaration-on-the-rights-of-indigenous-peoples.html

UNESCO (2005) *Convention on the Protection and Promotion of the Diversity of Cultural Expressions*. Available at: https://en.unesco.org/creativity/convention

Urban Waters Federal Partnership (2011) Available at: https://www.epa.gov/sites/production/files/2014-06/documents/uw-federal-partnership-report_v7al.pdf

Urry, John (2007) *Mobilities*, Cambridge: Polity Press

Van Dijck, Jose, Poell, Thomas and Martjin de Waal (2018) *The Platform Society: Public Values in a Connective World*, Abingdon: Oxford University Press

Van Dijck, Jose (2007) *Mediated Memories in the Digital Age*, Stanford, CA: Stanford University Press

Van House, N. (2011) 'Personal Photography, Digital Technologies and the Uses of the Visual' *Visual Studies* 26: 2, 125–34

Van Loon, Anne, Gleeson, Tom, Clark, Julian, van Dijk, Albert, Stahl, Kerstin, Hannaford, Jamie, Di Baldassarre, Giuliano, Teuling, Adriaan, Tallaksen, L. M., Uijlenhoet, Remko, Hannah, David, Sheffield, Justin, Svoboda, Mark, Verbeiren, Boud, Wagener, Thorsten, Rangecroft, Sally, Wanders, Niko and Henny Van Lanen (2016) 'Drought in the Anthropocene' *Nature Geoscience* 9: 2, 89–91

Vasterman, Peter (ed) (2018) *From Media Hype to Twitter Storm News Explosions and Their Impact on Issues, Crises, and Public Opinion*, Amsterdam: Amsterdam University Press

Viner, Brian (2015) 'Think We're Having a Heatwave?' *Daily Mail* on 6 July 2015

Wagner, John R. (ed) (2013) *The Social Life of Water*, New York: Berghahn Books

Waites, Ian (2018) 'Learning to Say "Phew" Instead of "Brr": Social and Cultural Change during the British Summer of 1976' in Endfield Georgina H. and Lucy Veale (eds) *Cultural Histories, Memories and Extreme Weather: A Historical Geography*, London: Routledge, pp. 16–33

Wainright, Martin (2007) 'The Great Floods of 1947' *The Guardian* newspaper, 25 July 2007

Wang, Susie, Leviston, Zoe, Hurlstone, Mark, Lawrence, Carmen, Walker, Iain and Diane Raines Ward (2018) 'Emotions Predict Policy Support: Why It Matters How People Feel about Climate Change' *Global Environment Change* 50, 25–40

Ward, Diane Raines (2003) *Drought, Flood, Folly and the Politics of Thirst*, New York: Riverhead Books

Wasik, Bill (2009) *And Then There's This: How Stories Live and Die in Viral Culture*, London: Penguin

Weick, Karl E. (1995) *Sensemaking in Organizations*, Thousand Oaks, CA: Sage

Westling, Emma, Surridge, Ben, Sharp, Liz and David Lerner (2014) 'Making Sense of Landscape Change: Long-Term Perceptions among Local Residents Following River Restoration' *Journal of Hydrology* 519: C, 2612–23

Whatmore, Sarah J. (2009) 'Mapping Knowledge Controversies: Science, Democracy and the Redistribution of Expertise' *Progress in Human Geography* 33, 587–98

Wheatley, Helen (2016) *Spectacular Television: Exploring Televisual Pleasure*, London: I.B. Tauris

Whitehead, Frederika (2017) *Water Wars: Fight to the Last Drop*, London: Curious Reads

Whitmarsh, L. (2008) 'Are Flood Victims More Concerned about Climate Change than Other People? The Role of Direct Experience in Risk Perception and Behavioural Response' *Journal of Risk Research* 11: 3, 351–74

Wild, T. C., Bernet, J. F., Westling E. L. and D. N. Lerner (2011) 'Deculverting: Reviewing the Evidence on the "Daylighting" and Restoration of Culverted Rivers' *Water and Environment Journal* 25, 412–21

Wiles, J., Rosenberg, M. and R. Kearns (2005) 'Narrative Analysis as a Strategy for Understanding Interview Talk in Geographic Research' *Area* 37: 1, 89–99

Wilhite, Donald A. and Olga Vanyarkho (2000) 'Drought: Pervasive Impacts of a Creeping Phenomenon, in Donald A. Wilhite (ed) *Drought: A Global Assessment, Vol. I*, London: Routledge, pp. 245–55

Williams, Raymond (1989 [1958]) 'Culture Is Ordinary' in Raymond Williams (ed) *Resources of Hope: Culture, Democracy, Socialism*, London: Verso, pp. 3–14

Williamson, Rosemary (2012) 'Breeding Them Tough North of the Border: Resilience and Heroism as Rhetorical Responses to the 2011 Queensland Floods' *Social Alternatives Special Issue: Disaster Dialogues: Representations of Catastrophe in Word and Image* 31: 3, 33–8

Wilson, Julie A. and Emily Chivers Yochim (2017) *Mothering through Precarity: Women's Work and Digital Media*, Durham, NC: Duke University Press

Withington, John (2013) *Flood: Nature and Culture*, London: Reaktion Books

Worcman, Karen and Joanne Garde-Hansen (2016) *Social Memory Technology: Theory, Practice, Action*, New York: Routledge

Workman, James (2009) *Heart of Dryness: How the Last Bushmen Can Help Us Endure the Coming Age of Permanent Drought*, New York: Walker Publishing

WWDR4 (2012) 'Managing Water under Uncertainty and Risk: The United Nations World Water Development Programme'. Available at: http://www.unesco.org/new/en/natural-sciences/environment/water/wwap/wwdr/wwdr4-2012/

Wynne, Bryan (2010) 'Strange Weather, Again: Climate Science as Political Art' *Theory, Culture and Society* 27: 2–3, 289–305

Zeisley-Vralsted, Dorothy (2015) *Rivers, Memory, and Nation-Building*, New York: Berghahn Books

Zizek, Slavoj (1996) 'I Hear You with My Eyes' in Renata Salaci and Slavoj Zizek (eds) *Gaze and Voice as Love Objects*, Durham, NC: Duke University Press, pp. 90–128

Index

Boldface locators indicate figures; locators followed by "n." indicate endnotes

Aaranovitch, Ben 143
Aarhus Convention 25 n.5
Abers, Naerea 165 n.16
accidental archives 166, 169, 174
Adger, Neil W. 42
administrations of memory 119–20
After Nature: Entangled Worlds (Jones) 2
'agenda-setting' 33 n.22, 42 n.8, 61,
    63 n.14
Ahearne, Jeremy 132 n.10
Alberta tar sands 34 n.23
Allam, Lorena 96 n.6
Allen, Matthew 87
Allett, Nicola 157 n.2
Alleyne, Brian 76
Allon, Fiona 34, 62, 90, 136
amphibious anthropology 7, 134, 134 n.17
Amrith, Sunil 90 n.19
Anderson, Deb 94, 94 n.2
Anderson, Mark 107–9
Andrews, Molly 76
Ang, Ien 120
Anishinaabe First Nation community of
    Canada 13
Annin, Peter 15
anonymity 46 n.15
Anthropocene 2, 137
    approach 106
    cultural memories of 11, 27
    mass media communication 27
    memory cultures 27
Appadurai, Arjun 58 n.4
The Archers (BBC Radio 4) 33, 33 n.20
archives 2, 9, 11, 17, 25, 41, 43, 48, 57–8,
    57 n.2, 86, 89, 95, 112–13, 116, 123,
    123 n.22, 129, 145, 148–9, 152 n.21,
    156. See also broadcast archives
    accidental 166, 169, 174
    BBC (see BBC Written Archives)

digital 120 n.16
    environmental consciousness 51
    evidence in 99, 104
    as Flood Memories 120 n.16
    Gloucestershire (see Gloucestershire
        Archives (Gloucestershire
        Heritage Hub))
    ITN 99, 104
    MACE 57 n.2, 99
    modalities of 124
    Pathe News 101, 101 n.15
    projects, memory 138 n.26
    of social networking 155
    Southern TV Archive 99
Ardern, Jacinda 136 n.22
Armitage, David 136, 137 n.23
Arnold, Matthew 135
arts and humanities centred
        interdisciplinary research (AHIR) 57
Arts and Humanities Research Council
    (AHRC), projects funded 3 n.3, 5
    n.11, 5 n.12, 6, 15, 17 n.32, 42 n.9,
    49 n.19, 138, 147 n.17
Ashdown, Isabel 100–1
Ashley, Richard 87 n.16
Ashton, Kevin 161 n.6
Askins, Kye 127 n.1
Assmann, Aleida 122, 138
Assmann, Jan 58
Atkinson, David 6
Attenborough, David 78–9, 78 n.5
Audience Evolution (Napoli) 33
audiences 2–4, 10, 18, 25, 27, 32, 35, 41,
    58, 59 n.7, 60, 63, 75 n.1, 81, 86,
    96, 106, 121–2, 127, 145, 148, 151,
    156, 160
    archival evidence 48–9, 49 n.21, 52
    community-based conversations 57
    consume water messages 37 n.26

of cultural taste 79, 79 n.8
environmental issues 39
media, use 64 n.19, 70 n.28, 75 n.1,
      115, 118 n.13, 164
memories 33, 40, 43
as producers 26, 43, 122
public as media-literate 68
social media 59, 70, 70 n.28, 79
under-served 9, 9 n.24, 111 n.1
water-related 'events' 24
Auge, Marc 172
Australian drought (1914) 10, 91, 94, 94 n.3
Authorised Heritage Discourse (AHD)
      89 n.17

Baake, Ken 93–4, 118
Bacallao-Pino, Lazaro M. 32
Bachelard, Gaston 7
Bakker, Karen 68, 68 n.24
Balch, Oliver 28 n.10
Ball, Philip 135
Bang and Olufsen Company 78
Barard, Karen 10, 173, 173 n.2
Barker, Peggy 39 n.1
Barnett, Cynthia 15 n.30
Barr, Damian 11 n.25
Barrett, Jeff 15 n.30
Barrier, Thames 48 n.18
Barthes, Roland 75
'basic needs' approach 68, 68 n.24
'bath-tub model' 137 n.24
BBC Written Archives 39 n.1, 40, 40
      n.3, 40 n.4, 49, 49 n.21, 51 n.25,
      52–4, 99
Be Water Smart (Thames Water) 71
Bear, Christopher 89 n.18
Beebeejaun, Y. 59 n.6
'being in the world' 84
Bell, David 12, 152
Bell, James 40 n.3
Bennett, Oliver 4, 4 n.9, 130, 131, 131 n.9
Berryman, Colin 67
Bhattacharya, Shaoni 97 n.10
The Big Blue Map of Bristol project 142 n.3
*Big Data: A Revolution That Will
      Transform How We Live, Work
      and Think* (Mayer-Schonberger &
      Cukier) 165 n.14
'Big Water' 62

Biggs, Reinette 167
*The Black Atlantic: Modernity and Double
      Consciousness* (Gilroy) 7 n.17
'the Blitz spirit' 44, 44 n.11, 46, 49, 100
The Blob (Yeaworth) 143 n.9
Blue Gold: World Water Wars (Bozzo) 37
      n.26
Blue Humanities Network 6 n.14
blue planet 6–7, 7 n.17, 9, 23 n.2
Blue Planet (Attenborough) 3, 78–9, 78 n.7
Blue Planet II (Attenborough) 78–9, 79 n.8
Blue Planet II Live 79
Blue Revolution 136
blue-washing 137
*Bodies of Water: Posthuman Feminist
      Phenomenology* (Neimanis) 133
*The Body and Physical Difference:
      Discourses of Disability* (Mitchell &
      Snyder) 103–4
Bohensky, Erin 163
Bolsonaro, Jair 136 n.22
bookshop display, Leamington Spa 80
Bower, Joseph 160, 160 n.5
Box of Broadcasts 99 n.13
Boyer, Dominic 16, 29 n.13
Boym, Svetlana 96 n.8
Bozzo, Sam 37 n.26
Brace, Catherine 82, 172
Bradshaw, Elizabeth 48 n.17
Brahmaputra River 164 n.11
Brazil 162–3, 172
      climate change prioritization 141
      cultural policy 140
      digital Museu da Pessoa 118, 143, 143
            n.6, 148
      drought of 1915 108
      environmental policy 136 n.22
      flood marks 164
      flood-risk in 165, 167
      Fundao dam disaster (2015) 141,
            141 n.1
      greening 141
      hydro-citizenship 142
      intangible cultural heritage 91
      Manuelzao Project 133, 133 n.14, 144
      memory archive projects 138 n.26
      Narratives of Water (2017–18) 86
      participatory decision-making 165,
            165 n.16

water projects 112
water scarcity 134
Waterproofing Data (2018–20) project
    121, 121 n.19
Brazil Geological Survey (CPRM) 165,
    167 n.17
Brecht, Berthold 155
Breuste, J. 16
Bricks and Water policy (UK) 162 n.9
Briggs, Asa 49–50, 49 n.22, 50 n.23
Brisbane Floods of 2011 5, 117 n.12, 163
Britain's Wildest Weather (Channel 4
    2014–2018) 78 n.7
British Universities and Colleges Video
    Council (BUFVC) 99 n.13
broadcast archives 26, 57, 57 n.2, 95, 99,
    105, 112, 123 n.22, 149, 152 n.21.
    See also BBC Written Archives
    communicative memory 58
    daylighting 42–3, 55, 58
    environmental consciousness 51
    flood memories of 1947 44–5
    flooded nations connected through
        media 53
    media as early warning system 52–3
    media producers as relief/compassion/
        care 53–4
    as memory agent 48
Brookey, Robert 81
Bruns, Axel 67 n.23
Bryant, Lia 84, 94, 127, 128 n.5
Bubbles: Spheres 1 Microspherology
    (Sloterdijk) 76–7
Buck, Gayle 158 n.3
Budds, Jessica 64
Bull, Jacob 111
Bulloch, John 15
Burgess, Jean E. 67 n.23
Burnham-Fink, Michael 158
Button, Geoffrey V 171

California drought (2015) 5, 5 n.10, 35, 69,
    69 n.27, 75, 172 n.1
campaigns 23 n.2, 61–3, 63 n.13, 63 n.14,
    63 n.15, 64 n.17, 142
Canal & River Trust (UK) 12, 130 n.8
canal and river system 130 n.8
Cap, Piotr 16

*Care for the Future-Connected
    Communities* 147 n.17
Castells, Manuel 30 n.15
Caton, S. C. 4, 18, 18 n.33, 23
Caughie, John 33
*Caught by the River: A Collection of Words
    on Water* (Barrett & Turner) 15 n.30
celebrity 24 n.4, 28, 34, 69 n.27, 135
Centre for Ecology and Hydrology (CEH),
    UK 12, 49 n.21, 139 n.28
Centres of Expertise and Civil Response 155
*Changing Places, Changing Lives – One
    Green Step at a Time* 151
Chapman, A. 34
Chappels, Heather 84 n.14
Chen, C. 7, 35–6
Cheong-gye Cheon restoration project
    146 n.15
China 31
    floods (1931) 155
    floods (1920s), covering 50
    social media platforms 66 n.21
Chivers Yochim, Emily 84 n.13
Christensen, Clayton M. 160, 160 n.5
cinema and water 24, 24 n.3, 69, 78, 78
    n.5, 137, 143 n.9
The Citizen (newspaper) 47
citizenship 5, 9, 18, 24, 29, 33–4, 40, 63,
    69, 75, 83, 87, 91 n.21, 140, 145,
    153, 166. *See also* hydro-citizenship
civil society responsibility 162
Clandinin, D. Jean 58
Clark, Wilma 59
Clarke, Gillian 15 n.30
climate change 1 n.1, 3, 3 n.3, 6 n.13,
    15–16, 26, 28, 39, 48 n.18, 51, 54,
    70 n.29, 79–82, 87, 90, 96 n.7, 107
    n.18, 114, 130 n.7, 153, 153 n.22,
    163, 172–3
    adaptation 158
    believers and deniers 51
    broadcast material 51, 51 n.24
    communication 26, 34, 55, 60, 70,
        150, 161
    'disconnected' young adults 165
    early warning infrastructure 49
    framings of 153, 153 n.22
    and mediated emotions 3 n.4, 34

memorial 29 n.13
mixed-media film 168 n.18
narratological approach 60
pre-date 124
prioritization (Brazil) 141
representations 107
shifting 70 n.29
water issues 9, 14
water scarcity 36
Climate Research Unit 64
'Climate Stories' 24 n.4
climate uncertainty 6 n.13, 9, 158
Coaffee, Jon 157, 165 n.15
Coates, Peter 5 n.11
Cohen, Bernard 33 n.22
Cohen, Margaret 6
Collins, Kevin 129
Commedia dell'arte 136 n.20
communication(s) 4, 7, 14–15, 15 n.28, 18,
    57, 71, 122, 124, 129–30, 141, 161,
    167–9, 173
  arts and humanities 158
  climate change 26, 34, 55, 60, 70, 150,
    161
  digital/digitized 1, 61, 61 n.11, 120
  drought 71, 95–6, 98, 100, 106, 109
  marketing and 2, 62, 70, 83
  Open Science 63
  protocol 121
  risk 24, 28, 120, 135, 139, 161
  science 8–9, 25, 27, 32, 62–4, 68, 72,
    81, 95 n.4, 103, 135, 150, 157
  social learning 129
  social media as 33, 58, 115, 171
  of water issues 24, 64, 76, 106, 115,
    120
  'water-related event' 25
  of water scarcity 3, 65, 100
Communities of memory and the problem
    of transmission (Pickering &
    Keightley) 124
communities on water issues 17, 25, 25
    n.5, 84, 134, 155, 162, 167–8
computer modelling (of floods) 168 n.18,
    169
conduits, media as 24–5, 142
Connerton, Paul 86, 112, 122
Connick, Sarah 128
Constant, Natasha 157

Contemporary Critical Discourse Studies
    (Hart & Cap) 16
*Contextualising Disaster: Catastrophes in
    Context* (Button & Schuller) 171
*Convention on the Protection and
    Promotion of the Diversity of
    Cultural Expressions* (UN) 128 n.4
*Convergence Culture: Where Old and New
    Media Collide* (Jenkins) 59
The Conversation 48 n.18
Cook, Bernie 40–1
Cook, James 90 n.20, 133, 133 n.15
Coronation broadcasts 50
corporate social responsibility (CSR) 2, 28,
    137, 144
Cotterill, Joseph 69
Cotton, Jacqui 25 n.7
Covid19 pandemic (2020) 11, 11 n.25, 70
    n.29, 173
Cox, Evelyn 99, 101–2
Craft, Aimee 13
Craps, Stef 11, 27
Critical Cultural Policy Studies: A Reader
    (Lewis & Miller) 131
Crow, Deserai 83
Crown Fountain, Chicago 135, 135 n.18
Cruz, Penelope 24 n.4
Cukier, Kenneth 165 n.14
cultural convergence 26
cultural economy approach 135, 138
cultural expressions 128 n.4, 135, 144 n.12
cultural/media production 3 n.3, 5–6, 9,
    33, 55, 111 n.1, 135, 151, 157, 174
  broadcast material 51, 51 n.24
  digitized communication 1, 1 n.1
  on environmental issues 28, 30
  memories 87, 128
  processes 36
cultural memory 11, 17, 32, 41, 43, 114,
    120, 128 n.5, 137, 141–2, 145–6,
    149–51, 159, 161
  Australian drought (1914) 10
  Brazilian digital Museu da Pessoa 118
  broadcast archives 58
  communicative memory and 58
  daylighting 42, 42 n.6
  research projects on 138
  studies 102, 138, 142
'cultural mesocosm' research 12

cultural place-maker 130
cultural policy 4–5, 4 n.9, 7, 36, 36 n.24,
        90, 107, 124, 127–31, 130 n.7, 135,
        140, 149, 172
    exploratory questions and scoping
        topics 13, 14
    implicit 132, 132 n.10
    policy transfer 12
Cultural Policy (Bell & Oakley) 12, 152
cultural values 4 n.5, 10, 12, 25, 29, 41, 50,
        90, 97, 111, 124, 127–31, 147–8,
        150, 172–4
    as lived-in experiences 26
    margins 131–5
    scenarios 136–40, 136 n.20
Cultures of Energy podcast (2015–19)
        29 n.13
Cultures of Optimism (Bennett) 4
cultures of water 1–2, 4, 4 n.7, 7, 11, 83,
        132, 133, 135, 142, 145
    flood heritage as 1
    manifestations of optimism 4
    multi-directional memory 136
Cusack, Tricia 144 n.13
Cutter, S L 42–3

da Cunha, Dilip 123 n.21
Dahlstrom, Michael F. 77
Daily Mail (newspaper) 97
Dalla Costa, Mariarosa 136
Darwish, Adel 15
Das Velhas (Old Lady's River) 132 n.12, 133
Davies, Ann 96 n.6
Davies, Caitlin 15 n.30
The Day After Tomorrow (Emmerich) 41
Day Zero 69, 107 n.18
Dayan, Daniel 31–2
daylighting 10, 58, 98, 142
    broadcast archives 42–3, 55
    hidden waterways 34, 144 n.12, 151
    Thames21's use of media 144
Dayrell, Carmen 141
de-canalization projects 144–6
de Sousa Santos, Boaventura 133–4
Deakin, Roger 15 n.30
The Declaration on the Rights of
        Indigenous Peoples 142 n.4
Defoe, Daniel 48
Degrossi, Lívia Castro 155 n.1

Del Toro, Guillermo 78 n.4
Deltares' 121 n.17
deluge 41, 42 n.7, 61 n.9, 70, 155, 174
digital age 16, 19, 39, 57, 63, 111, 171
    communicative memory 58
    social media flows 66 n.21
digital devices 27, 114 n.6, 166
digital forgetting 162
digital hydro-citizenship 61, 67–8, 70,
        77, 113
digital-hydro-public sphere 161
digital media 3, 25, 28, 48, 61 n.11, 90,
        111–12, 114, 121–2, 138, 143, 160,
        163–4
    'co-created memory' methods 119
    rapid response catchments 116 n.9
digital memory 61, 113–14, 119, 122, 143,
        163–4
digital remembering 162
digital stories 46, 46 n.14, 48, 67, 117,
        124, 149
digital storytelling movement 143
Disaster Resilience: An Integrated Approach
        (Paton & Johnston) 159 n.4
*Disaster Writing: The Cultural Politics
        of Catastrophe in Latin America*
        (Anderson) 107
disruptive events 8 n.23, 31, 31 n.16, 103,
        120, 160
diverse situations of water 36
'dockside kitsch' landscapes 6
Dolphijn, Rick 173
Donnellan, Philip 40 n.3
double season droughts (1933–4, 1975–6,
        1995–6) 109 n.21
Downpour! (game) 116, 116 n.10
Downriver (Sinclair) 143
*Downstream: A History and Celebration
        of Swimming the River Thames*
        (Davies) 15 n.30
drought 1, 3, 3 n.3, 9, 12, 15 n.29, 17, 26,
        30–3, 57–8, 62 n.12, 91, 106, 109
        n.21, 130, 137, 139, 151–3, 157, 159,
        161, 167, 173–4. *See also* narratives
        of drought
    Australian (1914) 10, 91, 94, 94 n.3
    California 5, 5 n.10, 35, 69, 69 n.27, 75,
        106, 172 n.1
    environmental adaptation 35

liveliness 71–2
　as masculine 93–5
　media archives 2
　memories 14, 102, 105, 169
　moral drought 76, 108
　and national culture/development 108
　plan 4, 71, 109
　policymakers 93, 109
　prediction 109
　risk and representation 24, 34, 70, 84,
　　96 n.7, 99, 104, 107–8, 110, 128,
　　139, 168, 174
　spectacles of hot weather 104–6
　Syria (2006–11) 70 n.28
　triggers zone 109
　UK drought (1976) (*see* UK drought
　　(1976))
　water management and plan 109
Drought: A Novel (Fraser) 1 n.2
Drought 2012: An Inside Out Special (TV
　　Programme) 99 n.12, 99 n.13
'Drought in the Anthropocene' 106
*Drought Plan 2014: Our plan for managing
　　water supply and demand during
　　drought* 109 n.21
Drought Risk and You (DRY) project
　　(2014–18) 71, 71 n.30, 86, 95, 95
　　n.4, 96 n.7, 97–8, 108, 147 n.17, 151
Drought Shame app 69
drought shaming 61, 68–72, 69 n.25, 69
　　n.27, 76, 150
Dry and Drier in West Texas 62 n.12
dry planet 109

early warning systems 48 n.18, 52–3, 95,
　　107, 155, 167
Earth Policy Institute (EPI) 97, 97 n.10
East Coast Flood (1953) 10, 53
East Coast Storm surge (1953) 40, 42, 42
　　n.8, 48–9, 48 n.18, 49 n.22, 53, 55,
　　100
East Coast Storm surge (2013) 42
Eating the Ocean (Probyn) 6
*Echoes from the Poisoned Well: Global
　　Memories of Environmental Justice*
　　(Washington) 112
eco-ethics 35
eco-media critique 27
ecofeminist waterways 132

'Ecological Ethics and Media Technology'
　　(Maxwell & Miller) 28
ecological suicide 35
ecology 5, 16, 30, 32, 37 n.26, 41–2, 41 n.5,
　　63–4, 82, 95 n.4, 106, 112, 117, 149,
　　152, 171, 173
Ecology and Society 113 n.4
Economic and Social Research Council
　　(ESRC) 42, 147 n.17, 150
ecosystem 4 n.5, 4 n.6, 9, 30, 58 n.3,
　　128, 145
Eden, Sally 89 n.18
Edwardians 33 n.21
Edy, Jill A. 27 n.9
electronic monuments 32
Elizabeth II, Queen 50
Emmerich, Roland 41
emo-techno-ecology 82
Emotion Online (Garde-Hansen & Gorton)
　　76, 82
emotions 1, 3 n.4, 7–8, 24, 31, 44, 48 n.16,
　　65, 106–7, 119, 129, 139, 158
　emotional geographies 89
　local and global 82
　mediated 34
　online 68–72, 75
　value 166
Endeavour 90 n.20, 133 n.15
Endfield, Georgina 5 n.11, 17 n.32, 49
　　n.19, 59
Endtar-Wada, Joanna 70
Endurance: Australian Stories of Drought
　　(Anderson) 94 n.2
Engle, Nathan L. 165 n.16
Environment Agency (UK) 12, 25 n.7,
　　46 n.14, 48, 85, 109, 111, 144, 145
　　n.14, 156
environmental policy 5, 9, 27, 54, 124,
　　135, 136 n.22
environmental studies 26, 29
environments (platform) 58 n.3
environments, media as 24–5
Epistemologies of the South (de Sousa
　　Santos) 133–4
Equal Pay Act (1970) 102
ESRI 116
European Environment Agency 4 n.6
European Union Water Framework
　　Directive 5, 25 n.5, 142

Evans, Luke 152
Evernden, Neil 30
exceptionally low storage 109
extreme weather 3, 3 n.3, 5, 5 n.12, 14, 28
    n.12, 52, 59, 68, 82, 89, 99, 124, 172
    in climate change 41
    events 6 n.13, 81, 137
    mediation 26
    metaphorization of social media 61
Extreme Weather and Global Media
    (Leyda & Negra) 34, 104

*The Fabric of Space: Water, Modernity
    and the Urban Imagination* (Gandy)
    15 n.30
Fairbanks, Eve 107 n.18
fans 2, 28, 59 n.5, 61, 81
*Fans, Bloggers, and Gamers: Exploring
    Participatory Culture* (Jenkins)
    59 n.5
Feldman, David Lewis 5
Ferguson, Benjamin 172 n.1
Ferguson, Daryl 91
Ferguson, Tristan 6
Figgener, Christine 23 n.2
filter bubbles 83
The Fish Ladder: A Journey Upstream
    (Norbury) 15 n.30
Fisher, Mark 82
Flagg, Charlie 93
flash flooding 116, 116 n.8, 116 n.9, 130 n.7
Fleming, Lionel 40, 40 n.3
The Flood 50
Flood: Nature and Culture (Withington) 155
flood community 8, 11, 44, 117, 118 n.13,
    118 n.14, 119, 122, 156
flood events 17, 39, 41–3, 48–9, 49 n.21,
    54, 61, 64 n.19, 87, 117–18, 118
    n.13, 123–4, 155, 162, 166
    catastrophic 171
    Downpour! (game) 116, 116 n.10
    early warning system 52
    memories of 50–1, 116
flood fatigue 123
flood memories (1947 & 2007) 11, 27,
    40–1, 43 n.10, 46 n.13, 86, 123, 156,
    159–60, 168–9
    community group sharing 46
    economies of 88

exhibition 121
flood marks 45
    to flood resilience 48
    media producers 44
    memorial programme 50
    memory agents 48
    as metadata 114–17, 159, 164–5
    monumentalizing/memorialization 44,
        116, 120, 166
    risk and resilience 44, 46 n.14
    sources at Gloucestershire Heritage
        Hub 47
    Winter Flood 43
flood memory app 82, 85–6
    beta version of 117
    digital media and memory work
        117–19
    flood memory as metadata 114–17
    water stakeholders, flood memories to
        119–22
    'watery sense of purpose' 111–14
*Flood of Images: Media, Memory and
    Hurricane Katrina* (Cook) 40
flood-place-memory-story 166
flood protocols 28 n.12
flood-risk communication, problems 120
flood-risk mitigation strategies 111 n.1
'Flood Warning Systems' 52
floodies (Facebook groups) 8 n.23
flooding/floods 3, 3 n.3, 5, 9, 17, 26–7,
    27, 31, 54, 87, 113 n.3, 151, 153,
    163, 174
    British media attention 43
    catastrophic 87, 155, 171
    cinematic treatments 24, 24 n.3
    in commemorative culture 6
    disaster narrative 121
    flood marks 86–8, 115, 164
    home video, evidence 87, 87 n.16, 88
    images 4, 23, 23 n.2, 40, 42, 44, 87, 155
    lost knowledge 163
    media archives 2, 44, 57
    media stories of 43
    memorials/memorialization 5, 44, 116,
        120
    'political economy' approach 114
    repeat experience/anecdote 156
    representations 3, 5, 24, 34, 40–1, 49,
        117, 117 n.12

risk (management) and resilience 18, 42, 42 n.9, 46 n.14, 61, 87, 89, 113, 115–16, 116 n.8, 120–1, 123 n.22, 124, 155–6, 159, 162, 165–6, 168–9
sensing (*see* sensing floods)
sub/urban-focused visual schema 113 n.3
through media 39, 53
Floodtags 121 n.17
Florida, Richard 36 n.24
Flow: For the Love of Water (Salina) 37 n.26
flux 7, 7 n.20
and flow 7–8, 7 n.21, 123
material metaphors of 7
forgetting. *See* remembering-forgetting (water cultures)
Foucault, Michel 62
framing 33 n.22, 34, 162
big floods 39–42
climate change 153, 153 n.22
Severn Trent's approach 65
Fraser, Ronald 1 n.2
French NGO Solidarites International 63 n.15
From Media Hype to Twitterstorm (Vasterman) 61 n.9
Frozen Planet 78 n.7
Fundao dam disaster in 2015 141 n.1
Furedi, Frank 44 n.12
future flows 139, 139 n.28
The Future of Heritage as Climates Change: Loss, Adaptation and Creativity (Harvey & Perry) 130 n.7, 138

Gandy, Matthew 15 n.30
Gane, Jo 144
Garde-Hansen, Joanne 76, 82
Garnham, Bridget 84, 94
Geiß, Stefan 58, 60
Genette, Gerard 75, 81
Genome 50
Geoghegan, Hillary 82, 172
George, Jodie 127, 128 n.5
Getty Images 57 n.2, 99, 105
Geyer, R. 78 n.6
Gilligan, Paula 104–5
Gilroy, Paul 7 n.17
GIS-enabled platforms 116
Glasauer, Herbert 16

Global Green Media Production Network 3 n.3
global media production 28
globalisation, scalar dynamics 58 n.4
Gloucestershire Archives (Gloucestershire Heritage Hub) 111
digital stories 46, 46 n.14, 48
as memory agent 48
sources at 47
workshops at 45–6
Gloucestershire Echo 42 n.7
Gonzales, Mike 15 n.30
GoodReads 15
Gorton, Kristyn 76, 82
Goulburn River (Australia) 147 n.16
Graham-Leigh, John 15
Grand, Ann 63
Gray, Jonathan 81
Gray, Sylvia 40 n.3
Great Drought of 1877–9 108
The Great Drought of 1976 (Cox) 99
The Great Gale 50
The Great Lakes Water Wars (Annin) 16
'a great news story' *vs.* 'a great ceremonial event' 31
The Great Water Risk: The Problems of Flooding 50
Green Revolution 2
The Green Studies Reader (Head) 6
Greene, Hugh Carleton 54
greening media 7, 31, 124
Greening the Media (Maxwell & Miller) 137 n.23
Greenpeace 23 n.2
greenwashing 28
Griffiths, H.M. 6
Griggs, Mary Beth 79
Grodach, Carl 146 n.15
Groundwork 144, 149, 151, 151 n.19, 161
The Guardian (newspaper) 28 n.10, 41, 43 n.10, 133 n.16
Gubbi, Jayavardhana 161
Gustafsson, Tommy 27

H_2O: *A Biography of Water* (Ball) 135
Hageman, Andrew 27
Haigh, Ivan 48 n.17
Halbwachs, Maurice 90
Hall, Alexander 10

Hall, Stuart 111
Hamburg flood 27, 162
Hamlin, Christopher 5
Handbook of Narrative Enquiry: Mapping a Methodology (Clandinin) 58
Hard Landings: Memory, Place and Migration 6 n.14
Hard Weather: Flood and Inundations of 1953 50
Hargreaves, Jack 98 n.11
Hart, Christopher 16
Harvard Business Review 160 n.5
Harvey, Colin 75 n.1
Harvey, David 130 n.7, 138
Hassan, John 5
Hassid, Jonathan 66 n.21
Haughton, Graham 163
Head, Lesley M. 84 n.12
Head, Dominic 6
healthier urban waters 3 n.4
Heart of Dryness (Workman) 107
Heat Wave in London: Women's Summer Fashion (1976) | British Path 101
heatwave 3 n.3, 4, 95, 98, 100, 104–6, 109–10, 153, 174
   culture 102
   memories of 96–7
   nostalgic video 101
Heise, Ursula K. 30
Heritage, Memory and Identity (Isar) 120
Herve-Bazin, Celine 16, 23 n.1, 37 n.25, 63 n.13
Hicks, Anna 164 n.12
hidden waterways 34, 144 n.12
The History of Broadcasting in the United Kingdom: Sound and Vision (Briggs) 49
The History of the Provincial Press in England (Matthews) 49
A History of Water in Modern England and Wales (Hassan) 5
Hoare, Philip 15 n.30
Home News 49 n.22
Home Service 40, 40 n.3
Hoskins, Andrew 32, 44
Hossain, Amzad 13
House of Cards 27 n.9
Houston Flood Museum 87 n.15
Howard, F.E. 53

Howe, Cymene 16, 29 n.13
Hull Flood Project 162
Hulme, Mike 31, 153, 153 n.22
human and non-human memory 133
human ecology 30
Hurricane Harvey (2017) 3 n.3, 87 n.15
Hurricane Katrina (2005) 3 n.3, 5, 24 n.4, 34, 41, 41 n.5, 117 n.12
hurricane memory 138
'hurricane playbook' 41, 44
Hurricane Rita 24 n.4
Huwaei 58 n.3
hydro-citizenship 57, 61 n.11, 67, 70, 72, 77, 86, 113, 113 n.4, 139, 141–3, 149–50, 168
Hydro-citizenship Project (Jones) 6, 6 n.16, 15
hydrological globalization 16, 29, 29 n.13
Hypodermic Needle Model 33

Illingworth, Sam 9 n.24
Image/Music/Text (Barthes) 75
implicit cultural policymakers 132, 132 n.10
In the Wake of the Floods 50
The Independent (newspaper) 103
indigenous cultures 35
inequalities/crises/politics (water) 11, 13, 15, 27 n.9, 42, 64, 98, 131, 133
Informing Cultural Policy: The Research and Information Infrastructure (Schuster) 131
Ingersoll, Karin Amimoto 136 n.22
Ingold, Tim 113
Innes, Judith E. 128
Instructions for a Heatwave (O'Farrell) 100
intangible cultural heritage 86, 91, 137, 166
intercultural positionality of ocean 7 n.17
interdisciplinary research 11–12, 24, 29, 57, 147
intermediality 77, 77 n.3
International Conference on Software Engineering and Knowledge Engineering 155 n.1
International Flood Relief Show 54
International Journal of Social Research Methodology 157 n.2
internet of things 161, 161 n.6, 169
The Invading Sea 50

The Invention of Rivers (da Cunha) 123 n.21
Ioris, Antonio A. R. 130, 131
IPCC 136
irrigation policy and river practice 128 n.5
Isar, Raj 120
Ison, Raymond 158
ITN archive 99, 104

Jackson, Frank 103
Jacob, Ian 50
Jambeck, J. R. 78 n.6
Jansen, Bernard J. 67 n.23
Jenkins, Henry 59, 59 n.5
Jensen, Eric 135 n.19
Jha, Alok 15 n.30, 135
Johnson, Jennifer Lee 7 n.19
Johnston, David 159 n.4
Jones, Ivor 40 n.3
Jones, Katherine 13, 57 n.1
Jones, Michael 83
Jones, Owain 2, 6, 13, 15, 15 n.30, 57, 57 n.1, 61 n.11, 157
Jones, Roy 6
Jones and Staniland 132 n.11
Jordan's Story (Groundwork) 144, 149–53, 149 n.18
Just Water Challenge 62 n.12

Kääpä, Pietari 27
Kaempf, Charlotte 118
Katz, Elihu 31–2, 31 n.16
Keck, Margaret 165 n.16
Keightley, Emily 124, 133 n.13, 157 n.2
Kellner, Douglas 32 n.18
Kelly, Matthew 48 n.18
Kelton, Elmer 93–4
Kirtley, Chris 98 n.11
Klein, Naomi 25 n.6, 34 n.23
Knapton, O. 16
knowledge deficit model 157
Kok, Kasper 168
Krause, Don 59
Krause, Franz 7, 63, 134 n.17

Lambert, D. 7
Lambert, Joe 118 n.15, 143
Landström, Catharina 168
Lane, Stuart, N. 169
language, media as 24–5, 103

The Last Drop: The Politics of Water (Gonzales & Yanes) 15 n.30
Latour, Bruno 82
Lavau, Stephanie 147 n.16
Law, K. Lavender 78 n.6
Leitch, Ann 163
Lejano, Raul 60, 145
Lemos, Maria Carmen 165 n.16
levee syndrome 159, 159 n.4
Lewis, Justin 131
Leyda, Julia 34, 104
Leyshon, Michael 111
Life on Earth (Attenborough) 78–9
Lindsay, Jason 37 n.26
Linton, Jaimie 64
Linton, Jamie 5
liquid memories 148, 153, 172
Liquid Power: Contested Hydro-Modernities in Twentieth-Century Spain (Swyngedouw) 1 n.2
Living Flood Histories Network (McEwen) 6
Living Flood Histories project 42 n.9, 85
Living on the Edge 145 n.14
Lloyd, Mark 103
London's Lost Rivers (Talling) 144 n.11
London's Water Wars (Graham-Leigh) 15
longue duree 94, 94 n.1, 101, 103
Longwell, Scott A. 67 n.23
'1953 Lord Mayor's National Flood and Tempest Fund' 54
Lynmouth – A Year After 50
Lynmouth flood disaster (1952) 48, 49 n.21, 50–1, 53, 55

Macfarlane, Robert 137 n.25
MacIntyre, Alasdair 76
Mackintosh, Sophie 78 n.4
MacLeod, J. 7
Mahmoud, Mohammed 136
'Major Hydrological Events' timeline 49 n.21
mamasphere 84 n.13
Managing flooding: From a problem to an opportunity (Ashley) 87 n.16
Mannheim, Karl 124
Mantel, Hillary 156
Manuelzao Project 133, 144

Maori for Tuia 250 Encounters 90 n.20,
133, 133 n.15
Maori 'guardianship' 136 n.22
Maori-led campaign 142
March, Hug 158
Marinova, Dora 13
maritime/costal turn 6, 79
Martin, H. 160 n.5
Marx, Karl 134
mass media 7, 28, 32 n.18, 33 n.22, 34, 82
communication of Anthropocene 27
templates 25–6, 31–2
Mass Observation Archive Sussex 99 n.12
Massumi, Brian 82, 172
Materiality and Memory (Rigney) 166
Mathieson, Charlotte 6
Matthews, Rachel 49
Mauss, Marcel 18
Maxwell, Richard 28, 137 n.23
Mayer-Schönberger, Viktor 165 n.14
McConkey, James 130
McCormack, Michael 96
McEwen, Lindsey J. 6, 112, 169
McGuigan, Jim 64
Meadows, Daniel 118 n.15
Meadows, Donella (Dana) 129
The Meaning of Water (Strang) 131
media 11, 14–15, 15 n.28, 76, 159, 171,
174. *See also* communication(s);
mass media
archaeological approach 116
as bubbles 83, 87
and communication 18, 27–8, 31, 33,
37, 49, 86, 95 n.4, 127, 130, 138
and creativity 142, 147, 149
and cultural communication 72
as early warning system 52–3
flooded nations connected 53
forms of 24–6, 41, 69 n.27, 75 n.1,
119, 166
organizations 103
producers as relief/compassion/care
53–4
as social technology 86
storying of water (*see* story-ing water)
Media Archive for Central England
(MACE) 57 n.2, 99
media events 15 n.29, 40, 53, 75–6
'broadcast' approach 32

climate-related events 32
and disruptive mediated events
31–2
eco-challenge, necessity 35
global memory 32
as 'media spectacles' 32 n.18
newsworthy templates 33
social media conversations 32 n.17
tabloidization 34
media 'eventscape' 31, 58, 64
media infrastructures 15 n.29, 17, 25, 27,
31, 39, 42, 48–54, 59, 62, 64, 103,
113, 117, 145, 169
Media Lifecycle Solutions 152 n.21
media production. *See* cultural/media
production
media studies 1–2, 8, 19, 24, 26, 35, 87,
119, 134, 158, 173
and environmental issues 26–9
global media production 28
greening/blue-ing 3, 6, 31, 137, 137
n.23
liquid metaphors 29–30, 29 n.14
watery turn 3
media templates 3, 18, 23, 32, 70, 117
and Hurricane Katrina 41
mediating war 44
for water issues 31, 34
Welsh tourism's 149
media treatment 17
media tsunami 61
mediascape 32, 58, 58 n.4, 76, 123
'mediated memories' 42 n.8, 43, 45, 111,
113, 143, 151, 172, 174
mediations of water 2, 9–10, 15–19, 24, 32,
34, 68, 87, 132
memories 68, 90
personal or community 58
in Sao Paulo and UK 61 n.11
Mediterranean Sea 4
Meeting the Universe Halfway (Barard)
173 n.2
Mellor, Mary 132
memo-techno-ecology 42
memory activation 151
Memory and Political Change (Assmann &
Shortt) 138
memory landscape 163
Memory of the World project 130 n.7

memory studies 2, 19, 36, 89–90, 114, 130, 138, 142, 147–8, 153, 163, 164 n.10, 167, 171–3
Mentz, Steven 6
mesocosm research 12
the Met Office 28 n.12, 32 n.19, 52, 100
Methodological Challenges in Nature-Culture and Environmental History Research (Thorpe) 13
Meyrowitz, Joshua 24
Michael, Mike 81
Midweek Theatre: Still Flows the Flood 50
Milbrandt, Tara 69 n.27
Miller, Toby 28, 131, 137 n.23
'Minute' (2012) 63 n.15
Mississippi River 133, 136
Mitchell, David 103
Mitchell, Kyle 144 n.13
mnemonic imagination 124
The Mnemonic Imagination: Remembering as Creative Practice (Keightley & Pickering) 133 n.13
mobile app for water waste issues 61 n.10. *See also* flood memory app
moral economy of water 70
Moran, Joe 76 n.2
Morgan, Ruth 10
Morrison, Toni 132, 136
Muir, Pat 84 n.12
Multi-Story Water (2012–17) 86, 147 n.17
multi-year droughts 109
multidirectional memory 102, 136
multimedia journalism 62 n.12
multimodality 77, 77 n.3
Mulvey, Laura 101 n.14
Murray-Darling river system 91, 96
Murthy, Dhiraj 67 n.23
Museu da Pessoa (Museum of the Person) 118, 143, 143 n.6, 148
Museum of Water and Steam 143 n.10
Myerhoff, B. 58
Myers, T. A. 3 n.4

Napoli, Philip 33
Narrative Discourse: An Essay on Method (Genette) 75
narrative methods 59, 59 n.6, 106
narratives of drought 11–12, 93–5, 97–8, 101, 108–9

National Farmers Union 12
National Flood Forum 33
National Library of Scotland 57 n.2
Natural Environment Research Council (NERC) 24 n.4, 48 n.17
The Natural History of Memory project (Bond, Rapson and Crownshaw) 164 n.10
natural resources 3, 9, 25, 27–8, 35, 54, 68, 113, 134, 142, 149
The Natural Resources Policy 152
Natural Resources Wales (2016–17) 152
Negra, Diane 34, 41 n.5, 104
Negrete, Aquiles 139
Neimanis, Astrida 7, 133
The Neoliberal Self (McGuigan) 64
The Netherlands Thank the World 53
Nettley, Amy 168 n.18
Neverla, I 27
The New York Times (newspaper) 103 n.16
New Zealand 90, 90 n.20, 133, 133 n.15
Maori 'guardianship' 136 n.22
non-human rights to memory, river 142
News Commentary for Schools 53
News Talks Feature 40 n.3
Nietzsche, Friedrich 137
Noah 50, 81, 155–6
non-governmental organizations (NGOs) 4, 12, 16, 18, 33, 61, 63, 120, 130 n.7, 132, 136, 139 n.27, 144, 149–51, 153, 166, 169
non-representational theory 1, 25
Norbury, Katherine 15 n.30
Northwood, Bill 40 n.3
nostalgia, types 96, 96 n.8
Notley, Tanya 143
The Novel and the Sea (Cohen) 6
Nye, M. 168

Oakley, Kate 12, 152
The Ocean: The Great Tidal Surge of 1953 50
Oceanic Histories (Armitage) 137 n.23
Oceanic Memory: Islands, Ecologies, Peoples 6 n.14
Odih, Pamela 132
O'Farrell, Maggie 100–2
OFWAT 64, 64 n.18

Old and New Media after Katrina (Negra) 41 n.5
Oliver-Smith, Anthony 15 n.31
On Violence (Brecht) 155
Once upon a River (Setterfield) 80
Onesto, Chris 5, 5 n.10
online emotion 68–72, 75
online memory resource 149
online platforms (media convergence) 136
open media infrastructure 62, 64
Open Street Maps 116, 117
Orientalism (Said) 25 n.6
Orlove, B. 4, 18, 18 n.33, 23
O'Sullivan, Tim 7 n.21
Our Planet (2019) 79, 79 n.9
Out of Town (Hargreaves) 98 n.11
The Outstanding Universal Value of Petra, Jordan 130 n.7
'overwhelming and touching aid' 53

Pakistan Floods of 2010 5, 67 n.23, 117 n.12
Pamuk, Orhan 143, 143 n.8
Pandurang, Mala 77
Paratexts: Thresholds of Interpretation (Genette) 81
paratextual liquidity 81–3
Parikka, Jussi 116 n.11
Pariser, Eli 83 n.11
participatory flood cultures 157–9
Paschen, Jana-Axinja 158
Pasotti, Eleanora 34
Passport to Mainstreaming Gender in Water Programmes (United Nations) 109 n.19
Pathe News 101, 105, 105 n.17
Pathe News archive 101, 101 n.15
Paton, Douglas 159 n.4
Pearce, Cathryn 6–7
The Perception of the Environment: Essays on Livelihood, Dwelling and Skill (Ingold) 113
perceptions of water 35, 124, 129–30, 139, 142, 145, 152, 162
Perry, Jim 130 n.7, 138
Photovoice 59 n.5, 91 n.21, 158 n.3
Picken, Felicity 6
Pickering, Michael 124, 133 n.13, 157 n.2
Pini, Maria 157
Pitt Review 2008 162

Planet Earth series (NERC) 48 n.17
plastic pollution 3, 23 n.2, 78
plasticity 24
platform power 28, 58, 58 n.3
The Platform Society: Public Values in a Connective World (van Dijck) 58 n.3
Play for Today: London Is Drowning 50
Plensa, James 135
Plunder: Lynmouth Flood 1952 50
Poldark (television series) 3
The Political Ecology of Women, Water and Global Environmental Change (Buechler & Hanson) 17
The Politics of Urban Cultural Policy: Global Perspectives (Grodach & Silver) 146 n.15
Popular Science (magazine) 79
post-Covid recovery 165 n.15
The Power and the Water project 5 n.11, 147 n.17
Princess Mononoke (Miyazaki) 143 n.9
Probyn, Elspeth 6
productive mutual learning 172
Property and the right to water (Schmidt & Mitchell) 144 n.13
prosthetic memories 138
PROTection of European Cultural HEritage from GeO-hazards (PROTHEGO) project 130 n.7
public engagement 10, 29, 31, 42, 107, 144, 147
pulp and paper industry, water in 30
Putting the Water into Waterways: Water Resources Strategy 2015–2020 130 n.8

quality, water 57, 64 n.19, 135, 144, 171
Quigley, Cassie 158 n.3
Quinn, Audrey 70 n.28

Radio Gloucestershire 64 n.19
Radio Nederland 53
Radio Newsreel 49 n.22
Radio Times 50
Radstone, Susannah 6 n.14
Rain (Talbot) 3
Rain: A Natural and Cultural History (Barnett) 15 n.30

'rapid response catchment' 116, 116 n.8,
116 n.9
RCUK 147 n.17, 151
*Re-connecting Children with Water
Memories* project 151
re-territorialization process 119
Reading, Anna 48, 114, 114 n.5, 119, 143
real environment 12
real-time flood maps 121 n.17
A Recipe for Water (Clarke) 15 n.30
*Reconnecting People and Water: Public
Engagement and Sustainable Urban
Water Management* (Sharp) 34
reflective nostalgia 96 n.8
remembering-forgetting (water cultures)
2, 9, 11, 26, 32, 36, 51, 87, 89, 133,
135, 153, 159, 171–3. *See also*
strategic forgetting
flood memory app (*see* flood
memory app)
Hurricane Katrina 34
institutions 113
memo-techno-ecology 42
memory activation 151
online/digital spaces/places 114, 124,
162, 164
river's right 141–5
through media 48, 53, 75 n.1, 111, 130
UK drought (1976) 95–8, 105
reporting water
arts and humanities perspective 76
communication of science 62–3
leaks and leakiness 64
and media interviews 64–5
Research Council UK 147 n.17
restorative nostalgia 96 n.8
Rigney, Ann 166
Rio Doce 141, 141 n.1
Rios e Ruas (Rivers and Streets) initiative
project 144 n.12
riparian landlords 145 n.14
riparian media and culture 36, 61, 130–1,
145–9, 153
bubbles method 153, 154
collection-recollection-circulation
phases 147
cultural eco-system 145
cultural values of water 147
'digital storytelling-style' stories 150
dynamic memory reservoir 148

hidden rivers and streams 147
Ladywell fields, river regeneration
146
liminal or transitional zone 146
liquidity to memory 149
river memory apps 147
river revitalization 148
sense of place 148
watery environmental activism 149–50
RISE 66 n.22
*The Rise of the Network Society: The
Information Age* (Castells) 30 n.15
*Rising from the Depths: Utilising Marine
Cultural Heritage in East Africa* 6
River Avon (Bristol) 142 n.3
River Avon project 142 n.3
river-human relations 141
river memory 147–8, 153
River Ravensbourne (London) 144, 144
n.11, 146
river regeneration 124, 128
Ladywell fields 146
research, questions for future 154
urban 11, 14, 28
River Severn (Bristol) 71, 88, 151, 156
River Sherbourne (Coventry) 83, 141
n.2, 144
River Stour (England) 132
River Tamar (Cornwall) 168 n.18
River Thames (England) 143
cultural and media engagement 143 n.10
plastic pollution 139
Rivers of London (Aaranovitch) 143
river's right
corridors for sustainable living 141–2
non-human actor 143
questions and projects 144–5
Riverscapes and National identities
(Cusack) 144 n.13
Roberts, Carolyn 103
Roberts, Liz 157
Robinson, Sue 34
Roche, Jackie 70 n.28
Rodder, Simone 26
Roeser, Sabine 139
Room for the River/Ruimte voor der Rivier
strategy 162, 162 n.8
room for water 162–4
Rossiter, Ned 119
The Royal Statistical Society 78 n.6

Rundblad, G. 16, 64 n.19
Ruppert, Evelyn 114 n.6
Ryan, Marie-Laure 77 n.3

Said, Edward 25 n.6
Salina, Irena 37 n.26
Salisbury, E. 6
Salmon, Christian 127 n.2
Salvaggio, M. 91
San Diego County Water Authority
    61 n.10
Sandover, Rebecca 28 n.10
Santos, Brigida 106
Sayers, Paul 106
scapes of cultural flow 58, 58 n.4
scenarios/scenario-ing 97, 130, 135,
    136–40, 136 n.20, 158
  change factors 139, 139 n.27
  competency groups 168
  digital use, principles 169
  downscaling 168
  flood risk and resilience 166–7
  forecasting and backcasting 168
  plausibility 167–8
  probability 168
  uncertainty 158
  upscaling 168
Schenk, G. J. 15 n.31
Schmidt, Jeremy 144 n.13
Schofield, R. A. 40 n.3
Schuller, Mark 171
Schuster, J. Mark 131
science communicators 55, 59 n.7, 63
Sciences and the Humanities 10
The Sciences' Media Connection
    (Rodder) 26
The Sea Inside (Hoare) 15 n.30
sea-level rise 133, 133 n.16, 134 n.17, 137,
    137 n.24
Sea Narratives (Mathieson) 6
The Secret Life of Ice (BBC4) 17 n.32
Selsey, Valentine 40 n.3
Selwood, John H. 6
Sendai Framework for Disaster Risk
    Reduction (SFDRR) 165, 165 n.13
sense of place 113, 128 n.5, 148, 154
  amphibious 143
  dry 1, 14, 17, 26, 70, 86, 90, 95–6, 98,
    151, 166
  forested and mountainous 149

watery 1, 10–11, 16–17, 26, 29, 37,
    51, 70, 85–6, 90–1, 96 n.5, 98, 110,
    112–13, 140, 143, 146, 149–53,
    164–6, 169, 171, 174
sensing floods 124
  climate change communication 161–2
  digital banner at London 159, 160
  'hackable' innovations 160
  sensors 161
  social movements 161
sentiment analysis 114 n.7
Seoul Tomorrow policy 135
Setten, Gunhild 6
Setterfield, Diana 80
Seuils (Genette) 81
severe drought in 1959 109 n.21
Severn Trent Water 64–5, 64 n.18, 67
  slogan 65 n.20
  Twitter, use of 66–7, 71
Severn Trent Water Drought Plan 2014–19
    109, 109 n.20, 109 n.21
Shacklock, Zöe 51 n.25
The Shape of News Waves (Geib) 60
The Shape of Water (Del Toro) 78 n.4
Sharp, Liz 34, 136 n.21
Shiva, Vandana 15
Shortt, Linda 138
Shukman, David 162 n.7
Silver, Daniel 146 n.15
Sinclair, Iain 143
single year droughts (1921, 1984 and
    2003) 109 n.21
Sklair, Leslie 27
Skoric, Marko M. 69 n.27
Sloterdijk, Peter 76, 83, 154
smart media technologies 28
Smith, Benjamin 144 n.13
Smith, Laura Jane 89 n.17
Snagge, John 40 n.3
Snow Scenes: Exploring the Role of Place in
    Weather Memory 17 n.32
snowballing techniques 46 n.13
Snyder, Sharon 103
social learning 11, 43, 122, 129, 156,
    161, 165
social media 2–3, 15 n.28, 25, 27, 32 n.19,
    33, 37, 40, 55, 61 n.9, 61 n.11, 62
    n.12, 66 n.21, 66 n.22, 70, 72, 76,
    85, 95, 123–4, 144, 149, 151, 158,
    161, 164, 169, 171

campaigns 64 n.17
conversations 32 n.17, 66
crisis relief 121
drought-shaming 69
leakiness 68
mapping flood memory 41–2
and online emotion 75
rapid response catchments 116 n.9
right to forget 118 n.13
Severn Trent Water 71
storms of 57–61
for water issues 63, 115
'social memory technology' approach 148
social sciences 2, 5–6, 8, 24–5, 27, 29, 37,
    48, 51, 58, 62, 76, 107, 119, 127,
    129, 137, 139, 150, 157, 163, 173
socio-ecological resilience 156
Soechtig, Stephanie 37 n.26
Sofoulis, Zoë 34, 62, 90, 136
Solomon, Steven 135
Somerville, Margaret 91
Sony UK Technology Centre 152, 152 n.21
South Africa: How Cape Town beat the
    drought (Cotterill) 69
South East England drought (2006)
    84 n.14
Southern TV Archive 99
*Spaces of Experience and Horizons of
    Expectation: The Implications of
    Extreme Weather Events, Past,
    Present and Future* (project) 5 n.12
Spare Rib Magazine 102
*Spectacular Television: Exploring Televisual
    Pleasure* (Wheatley) 78
sphere theory 83–4
Spheres: Volume 1 Bubbles (Sloterdijk) 154
spreadability of media 33
Spy in the Snow (BBC1) 78 n.7
State of Natural Resources Report 152
'steam' metaphors 66 n.21
Stedman, Lis 62
Stehlik, Daniela 94
Stephens, C. 127 n.3
'still dominant cybertarian position' 28
storage 86–91, 109 n.20
Storm Chiara (2020) 3 n.3
The Storm (Penguin Classic, 2005)
    (Defoe) 48
Storm Dennis (2020) 3 n.3

Storm Doris 32 n.19
storm-naming 32 n.19
storms/storm surges 15 n.29, 33, 41–2,
    42 n.8, 48, 48 n.17, 76, 94 n.1, 137,
    174. *See also* East Coast Storm
    surge (1953)
    broadcast on radio/television 50
    cinematic treatments 24 n.3
    representations 24, 40
    of social media 57–61
    Twitter 17, 32 n.17
Story Center in California 118 n.15
The Story Group (Colorado) 62 n.12
story-ing water 35–6, 72, 75, 132–3, 161
    authorship and ownership, sense 75,
    75 n.1
    The Blue Planet 78
    bubbles 76–7, 83–6, 85, 153
    paratextual liquidity 81–3
    storage 86–91
    transmedia 75, 75 n.1
storyboard techniques 135, 135 n.19
storying of Wales 152
storytelling 2, 9–10, 18, 32, 34–5, 59, 59
    n.6, 76–7, 79, 83, 86, 89–90, 120,
    128, 131, 133, 144 n.11, 145, 153,
    158, 163–6, 171–2, 174
    digital 12, 85, 118 n.15, 119, 121, 129,
    143, 147–8, 150–1
    drought 98–104
    for exploratory purposes 167
    immaterial 114, 164
    integrative method 136
    for supporting decisions 168
Storytelling: Bewitching the Modern Mind
    (Salmon) 127 n.2
Stracher, Cameron 15
Strang, Veronica 63, 67, 131–2
*Strange Weather, Again: Climate Science
    as Political Art* (Wynne) 62
strategic forgetting 86, 96, 117, 122, 137
    factoids 97
    types 122, 122 n.20
Summer Floods of 2007 42, 43 n.10, 44, 49
    n.21, 64–5, 64 n.18, 85, 100, 111
Summer of '76 (Ashdown) 100
Sustainable Development Goal (SDG) 165
Sustainable Flood Memories (2010–13)
    project 42, 42 n.9, 85, 87 n.16,

100–1, 118, 120 n.16, 122, 147 n.17, 150, 159
Suzhou River (Hageman) 27
Swift, Graham 6
Swyngedouw, Eric 1 n.2, 8, 18 n.33
Syria's Climate Conflict (Quinn) 70 n.28

Talbot, Bryan 3
Talbot, Mary 3
Tales from the River Thames 143 n.10
Talling, Paul 144 n.11
Tam, Yee-Lok 27
Tapped (Soechtig & Lindsay) 37 n.26
Tempest (1953), media infrastructures 49
    broadcast material 50–1, 51 n.24
    environmental consciousness 51–2
Tencent 58 n.3, 171
The Texas Water Journal 93
Thames21 144
Thames catchment 143
Thames Water 66 n.22, 71, 103
Thinking in Systems (Meadows) 129
Thinking Relationships through Water
    (Krause & Strang) 63
Thinking with Water (Chen, MacLeod &
    Neimanis) 7, 35
Thompson, Paul 143
Thompson, William 78 n.4
Thon, Jan-Noel 77 n.3
Thorpe, Jocelyn 13
Thrift, Nigel 89, 95
Thussu, Daya Kishan 7 n.20
*Tides and the Ocean: Water's Movement
    around the World, from Waves to
    Whirlpools* (Thompson) 78 n.4
The Time It Never Rained (Kelton) 93–4
*To Walk with Water: Young People and
    Urban River Cultural Values in
    Spon End, Coventry* (project)
    83 n.10, 151
The Torn Halves of Cultural Policy
    Research (Bennett) 130
Totally Thames Festival 143 n.10
Towards Hydro-citizenship project 147 n.17
traditional wisdom 35
transmedia events 32
transmedial waterscape 112
transnational media waterworlds 164–6
transrurality 127 n.1

Troubled Water 50
troubled waters 60
True Stories: East Coast Flood 1953 50
Trümper, Stephanie 27
Tsunami of 2004 5, 117 n.12
Tuohy, R. 127 n.3
Turner, Robin 15 n.30
Tvedt, Terje 16
Twitter 11 n.25, 62, 65–6, 85, 115, 161, 171
    ad hoc 'hashtag publics' 115
    drought-shaming on 61, 69
    DRY Project 71, 71 n.30
    hashtags 67 n.23, 72, 150
    non-media organizations 59–60
    'Remember 1976' news feed 96
    risks and questions 72
    Severn Trent Water 66–7
    storms 17, 32 n.17
Twittersphere 32 n.17
Twitterstorm 60, 60 n.8

UK drought (1976) 72, 95–8, 98 n.11,
    100, 109
    collective memory 100–1, 102, 105
    creeping paralysis 103
    crippling drought 103
    evidence in archives 99, 104
    imagined community 101
    representation of women 101–2
'UK flood' on Google Trends 114
UK Water Management 98
under-served audiences 9, 9 n.24, 111 n.1
Under the Skin (Glazer) 143 n.9
Underland (Macfarlane) 137 n.25
United Nations (UN) 25 n.5, 78, 109
    n.19, 165
United Nations Educational, Scientific
    and Cultural Organization
    (UNESCO) 25 n.5, 91, 128 n.4,
    130 n.7, 172
The United Nations Framework on
    Climate Change (COP24) panel 78
University of Warwick 130
    Blue Humanities Network 6 n.14
    On Flood and Ice: Hydrological
        Globalization and the Rise of Water
        29 n.13
    *Global Green Media Production
        Network* 3 n.3

*Unruly Waters: How Mountain Rivers and
    Monsoons Have Shaped South Asia's
    History* (Amrith) 90 n.19
urban 'drought-shaming' 69 n.27
Urban Ecology (Breuste) 16
urban water 1, 3 n.4, 13, 36, 145, 148–9,
    154, 172
Urry, John 7 n.20, 141
Ushahidi platform 117, 161
'utilities gamification' model 66 n.22

valuation of water 130–1
value-belief-norm theory 91
van der Tuin, Iris 173
van Dijck, José 58 n.3
Vanyarkho, Olga 103 n.16
Vasterman, Peter 61 n.9
Veale, Lucy 59
Venice in Peril 130 n.7
Vescovo, Victor 78 n.4
Victorians 33 n.21
Vimeo 62 n.12
Viner, Brian 97
visualisation, water uncertainties 139
visualisations through storyboarding 139
The Voice of the Past (Thompson) 143

Wainright, Martin 43 n.10
Walk with Water tour (Talking Birds)
    141 n.2
Walker, Gabrielle 17 n.32
Wang, Susie 34
Ward, Diane Raines 16
Warde, Paul 5 n.11
Washington, Hood 112
Wasik, Bill 115
water
    cultural policy making (*see* cultural
        policy)
    as culture 1, 6 (*see also* cultures of
        water)
    governance 11, 17–18, 34, 36, 83, 138,
        147–8
    imaginaries 27
    mediations of (*see* mediations of water)
    mobilization 8
    as natural resource approach 142
    as non-human actor 6, 9
    as online conversations 59

scarcity (*see* water scarcity)
slave trade/shipwreck
    commemoration 6
    as total social fact 18
    on trend 77–80
Water (Feldman) 5
*Water: The Epic Struggle for Wealth, Power
    and Civilization* (Solomon) 135
*Water and Dreams: An Essay on the
    Imagination of Matter* (Bachelard) 7
Water and Environment Journal 42 n.6
*Water and Rural Communities: Local
    Politics, Meaning and Place* (Bryant
    & George) 127
Water and Society (Tvedt) 16
Water and Wastewater Treatment 66 n.22
The Water Book (Jha) 15 n.30, 135
The Water Brothers 63
*Water Communication: Analysis of
    Strategies and Campaigns from the
    Water Sector* (Herve-Bazin) 16, 37
    n.25, 63 n.13
water companies 61, 64, 67–8, 68 n.24, 98,
    103, 120, 142, 150–1, 174. *See also*
    Severn Trent Water
    communication strategies 109
    Drought Plan 71
    online backrooms 109
    social media, use of 58, 62
water conservation 23 n.2, 93
water consumption 84 n.12, 84 n.14,
    172 n.1
The Water Cure (Mackintosh) 78 n.4
*Water Cycle Management for New
    Developments* project 136 n.21
water events 2, 17, 24–6, 32–4, 62, 65,
    81, 89, 159–60. *See also* drought;
    flooding/floods; media events;
    storms/storm surges
'Water Ink' (2011) 63 n.15
water issues 1 n.1, 1 n.2, 2, 8–9, 27, 32, 57,
    68, 75, 84, 89–90, 96 n.5, 145, 172,
    174. *See also* reporting water
    arts and sciences on 16
    communication 24, 62–4, 66, 106,
        120, 145
    communities on 17, 25, 25 n.5, 84, 134,
        155, 162, 167–8
    media templates 31, 34

mobile app for 61 n.10
online stories 61
public engagement 3, 68
science communication 62
social media for 63, 115
water campaigns 61
'Water Kills' (2007) 63 n.15
water law 13
water management 3, 16–18, 23–4, 32,
    59, 82–3, 91, 98, 112, 124, 131, 138,
    142, 158
    blueprint 13
    framing media 34
    Goulburn River, Australia 147 n.16
    micro-scales 67
    mobilization 165
    movements of NGOs 150
    online stories 61
    and plan for drought 109
    policy change 25
    in South Africa 107 n.18
    'systems-thinking' approach 129,
    129 n.6
water memory 148, 153, 164
water policymaking 8, 34, 91, 127–8,
    130 n.7
water regime 68
water research movements 7, 127
water scarcity 3–4, 9, 11–14, 16–17, 36, 42,
    61, 65, 68–9, 70, 72, 90, 100, 103,
    107, 110, 128, 172 n.1
    Brazilian community 134
    drought and 91, 98, 100
    education policy 99 n.13
    project 86
    representation of women 99, 102
    stories of 11, 62 n.12, 110
    women as carers 106
water stakeholders 15, 18, 68, 70, 94, 98,
    119–22, 123 n.22, 129, 136, 151,
    163, 174
water towers, re-use 135
water-users 10, 18, 61, 83 n.10, 93, 98, 128,
    132, 135, 166
'Water Wall' (2010) 63 n.15
*Water Wars: Coming Conflicts in
    the Middle East* (Bulloch &
    Darwish) 15

*Water Wars: Drought, Flood, Folly and the
    Politics of Thirst* (Ward) 15–16
Water Wars: Fight to the Last Drop
    (Whitehead) 16
*Water Wars: Privatization, Pollution and
    Profit* (Shiva) 15
The Water Wars (Stracher) 16
water weeks 23, 23 n.1
water zeitgeist, books 15, 15 n.30
'watercooler' moments 27 n.9
watercraft 89 n.18
Waterland (Swift) 6
*Waterlog: A Swimmer's Journey through
    Britain* (Deakin) 15 n.30
Waterproofing Data (2018–20) project 86,
    121 n.19
Waters, Christian 23 n.2
*Waters Rising: Making Art in Storm and
    Calm* 121 n.18
water's 'voice' 24 n.4
waterscape 18 n.33, 29, 59, 112
watersheds 18 n.33, 23, 68, 171
Watersheds in Marxist Ecofeminism (Odih)
    132
waterworlds 4, 4 n.5, 18, 63, 121, 162 n.9,
    164–6
'watery sense of place' 1, 10–11, 16–17,
    26, 29, 37, 51, 70, 85–6, 90–1, 96
    n.5, 98, 110, 112–13, 140, 143, 146,
    149–53, 164–6, 169, 171, 174
Waverley report 48 n.18
*Waves of Knowing: A Seascape
    Epistemology* 136 n.22
weather 17, 31–2, 39, 44, 54, 59, 66,
    86, 100, 120, 130 n.7, 133, 153
    n.22, 174
    dry 4, 97, 101, 109
    extreme 1, 3, 3 n.3, 5, 5 n.12, 6 n.13, 14,
    26, 28 n.12, 41, 52, 59, 61, 68, 76,
    81–2, 89, 124, 137, 172
    fire 4, 104
    hot 10, 71, 95–7, 103–6
    war 44, 46, 82, 99, 101
Weathered: Cultures of Climate (Hulme)
    153 n.22
'Weathering the Storm: TEMPEST and
    Engagement with the National
    Weather Memory' project 49 n.19

Weibo 58 n.3, 161, 171
Welsh community 149
Welsh Tourism 149, 152
What Is Media Archaeology? (Parikka)
    116 n.11
*What Is Water? The History of a Modern
    Abstraction* (Linton) 5
What Moves at the Margin (Morrison) 132
Whatmore, Sarah 29 n.14
Wheatley, Helen 78–9
When the Waters Came 50
Wild, T. C. 42 n.6
*Wild Weather UK: Winners and Losers
    2018* (ITV) 78 n.7
Wiles, J. 127 n.3
Wilhite, Donald A. 103 n.16
Williams, Raymond 7 n.21, 131 n.9, 134
Williamson, Rosemary 127 n.3

Wilson, Julie A. 84 n.13
Winter Flood (1947) 15 n.29, 42–3, 49,
    55, 100
Withington, John 155
Woman's Hour (BBC Radio 4) 39, 39 n.1
Women and Water in Global Literature
    132 n.11
Women's Movement 102
Workman, James 107, 107 n.18
The Wye, River (Wales) 151
Wynne, Bryan 62–3, 139, 143

Yanes, Marianella 15 n.30
Yorkshire Post 49
'Your Convenience Is Their Extinction'
    (Waters) 23 n.2

Zizek, Slavoj 49 n.20

www.ingramcontent.com/pod-product-compliance
Lightning Source LLC
Chambersburg PA
CBHW050431280326
41932CB00013BA/2068